Concrete
Formwork
Systems

Civil and Environmental Engineering

Additional Volumes in Production

Concrete Formwork Systems

Awad S. Hanna
University of Wisconsin—Madison
Madison, Wisconsin

CRC Press
Taylor & Francis Group
Boca Raton London New York

CRC Press is an imprint of the
Taylor & Francis Group, an **informa** business

First published 1999 by Marcel Dekker, Inc.

Published 2019 by CRC Press
Taylor & Francis Group
6000 Broken Sound Parkway NW, Suite 300
Boca Raton, FL 33487-2742

© 1999 by Taylor & Francis Group, LLC
CRC Press is an imprint of Taylor & Francis Group, an Informa business

First issued in paperback 2019

No claim to original U.S. Government works

ISBN 13: 978-0-367-44766-3 (pbk)
ISBN 13: 978-0-8247-0072-0 (hbk)

Visit the Taylor & Francis Web site at
http://www.taylorandfrancis.com

and the CRC Press Web site at
http://www.crcpress.com

Library of Congress Cataloging-in-Publication Data

Hanna, Awad S.
    Concrete formwork systems / by Awad S. Hanna.
        p. cm.—(Civil and environmental engineering series: vol. 2)
    Includes index.
    ISBN 0-8247-0072-4 (alk. paper)
        1. Concrete construction—Formwork.   I. Title.   II. Series.
TA382.44.H36   1998
624.1′834—dc21   98-37262
                                                                               CIP

Preface

Formwork development has paralleled the growth of concrete construction throughout the 20th century. In the last several decades formwork technology has become increasingly important in reducing overall costs, since the structural frame constitutes a large portion of the cost of a formwork system.

This book has three objectives. The first is to provide technical descriptions and evaluations of ten formwork systems that are currently used in concrete construction. The second is to serve as a tool to assist contractors in selecting the optimal formwork system. The third is to present the design criteria for conventional formwork for slabs and walls using the stress and the stress modification factors provided by the National Design Specifications (NDS) and the American Plywood Association (APA).

Following a comprehensive introductory chapter, five types of formwork systems for concrete slabs are presented in chapters 2–5. These are conventional wood forms, conventional metal forms, flying forms, the column-mounted shoring system, and tunnel forms. The last four chapters describe five types of formwork systems for concrete columns and walls: conventional wood forms, ganged forms, jump forms, slip forms, and self-raising forms. Particular consideration is given to topics such as system components, typical work cycles, productivity, and the advantages and disadvantages associated with the use of various systems.

The selection of a formwork system is a critical decision with very serious implications. Due consideration must be given to such factors as the system's productivity, safety, durability, and many other variables that may be specific to the site or job at hand. Chapters 5 and 9 provide a comparative analysis of forming systems for horizontal and vertical concrete work to facilitate the selection of the optimal forming system.

Existing formwork design literature is inconsistent with the design criteria for wood provided by the NDS and the APA. Chapters 3 and 7 provide a systematic approach for formwork design using the criteria of the American Concrete Institute committee 347-94, the NDS, and the APA. For international readers, metric conversion is provided in the Appendix.

This book is directed mainly toward construction management, construction engineering and management students, and concrete contractors. It may also serve as a useful text for a graduate course on concrete formwork, and should be useful for practicing engineers, architects, and researchers.

Awad S. Hanna

Contents

Contents

Acknowledgments

I gratefully acknowledge a number of individuals who were instrumental in some way in the completion of this book. I begin with my friends and colleagues at the University of Wisconsin–Madison. The support and encouragement of professors John Bollinger, Al Wortley, Jeff Russell, Dick Straub, and Gary Bubenzer will always be remembered. I would like also to thank my students for inspiring me to further explore the field of concrete formwork. Thanks also go to my student Alan Lau, who assisted in preparing the graphics.

Special thanks go to the many contractors who provided me with data and graphics. I would like to specifically thank the editorial team of Marcel Dekker, Inc. for their strong support.

I would like to convey my warmest thanks to my loving wife Paula and our son Rewais. It is impossible to describe how supportive Paula has been throughout the writing of this book. As I spent hundreds of long hours at my computer preparing this manuscript, her never-ending love and support inspired me to keep pushing on. Most importantly, she gave me the greatest gift of all, our son Rewais, who is truly my reason for living. My respect and love for both of them.

I also take this opportunity to thank my family for their undying love and devoted support over the years; in particular, my late parents, Soliman and Sofia Hanna, who taught me how to work hard and encouraged me to pursue my dreams. I would also like to thank my sisters, Evette, Mervat, Moura, Sonia, and Janette, and my brother, Maged, for supporting my endeavor. I also want to mention my special mother-in-law and father-in-law, Botros and Bernice Hemaya, who believe in me.

Finally, my thanks to Professors Jack Willenbrock and Victor Sanvido, my mentors—but most of all my friends. They have provided me with advice and with numerous other experiences over the years for which I will be forever grateful.

1

Concrete Formwork: An Introduction

1.1 CONCRETE CONSTRUCTION

A quality reinforced concrete structure offers many advantages over structures made with other building materials. Concrete is a durable material that reduces building maintenance costs and provides a longer service life. A concrete structure will reduce energy usage because of its mass and high resistance to thermal interchange. The use of concrete will lower insurance costs by virtue of its high resistance to fire. Buildings made of concrete are also more secure against theft and vandalism. Concrete floors and walls reduce the transfer of noise, yielding a quieter environment and happier occupants. Reinforced concrete possesses considerable strength for resisting seismic and wind loads. These factors and others make the selection of reinforced concrete an economical alternative.

1.2 CONCRETE FORMWORK

The construction of a concrete building requires formwork to support the slabs (horizontal formwork) as well as columns and walls (vertical formwork). The terms *concrete formwork* and *concrete form* carry the same meaning and are used interchangeably in this book. Formwork is defined as a temporary structure whose purpose is to provide support and containment for fresh concrete until it can support itself. It molds the concrete to the desired shape and size, and controls its position and alignment. Concrete forms are engineered structures that are required to support loads such as fresh concrete, construction materials, equipment, workers, various impacts, and sometimes wind. The forms must support all the applied loads without collapse or excessive deflection.

1.2.1   Formwork System

A formwork system is defined as "the total system of support for freshly placed concrete including the mold or sheathing which contacts the concrete as well as supporting members, hardware, and necessary bracing." Formwork system development has paralleled the growth of concrete construction throughout the twentieth century. As concrete has come of age and been assigned increasingly significant structural tasks, formwork builders have had to keep pace. Form designers and builders are becoming increasingly aware of the need to keep abreast of technological advancements in other materials fields in order to develop creative innovations that are required to maintain quality and economy in the face of new formwork challenges.

Formwork was once built in place, used once, and subsequently wrecked. The trend today, however, is toward increasing prefabrication, assembly in large units, erection by mechanical means, and continuing reuse of forms. These developments are in keeping with the increasing mechanization of production in construction sites and other fields.

1.3   FORMWORK ECONOMY AND SIGNIFICANCE

Formwork is the largest cost component for a typical multistory reinforced concrete building. Formwork cost accounts for 40 to 60 percent of the cost of the concrete frame and for approximately 10 percent of the total building cost. Figure 1.1a, b presents a breakdown of different cost categories for conventional concrete slab and wall formwork. A large proportion of the cost of conventional formwork is related to formwork labor costs. Significant cost saving could be achieved by reducing labor costs.

Formwork costs are not the only significant component of

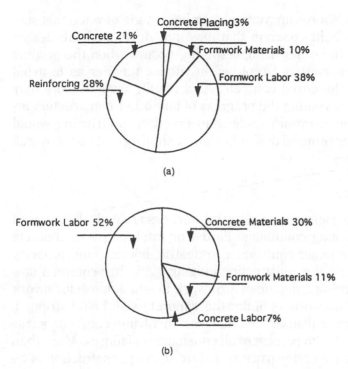

Figure 1.1 Distribution of costs for cast-in-place concrete slab wall: (a) slab; (b) wall.

the formwork life cycle. Other important aspects of the formwork operation include speed, safety, and quality.

1.3.1 Speed

Speed of construction is defined as the rate in which concrete building is raised and can be expressed in terms of number of floors erected per week or months. Speed of construction can be also measured in terms of inches or millimeters of concrete poured per hour. Formwork operations can control the pace of construction projects. Formwork is typically supported by several levels of shores and reshores that carry the loads until the concrete gains enough strength to support its own weight and all other externally applied loads. Shores are vertical members made of wood that support recently built concrete that have not developed full design strength. On the other hand, reshoring occurs when the original shoring is removed and replaced in such a manner as to avoid deflection of the cured concrete. As a result, several floors may be blocked, preventing the progress of any other construction activities. Faster formwork cycle from erection to stripping would allow for faster removal of shoring and reshoring and faster overall project progress.

1.3.2   Safety

Formwork operations are risky, and workers are typically exposed to unsafe working conditions. Partial or total failure of concrete formwork is a major contributor to deaths, injuries, and property damages within the construction industry. Another common hazard occurs during stripping of formwork in which loose formwork elements fall on workers under the concrete slab being stripped.

Structural collapses and failures involving concrete structures account for 25 percent of all construction failures. More than 50 percent of concrete structure failure during construction is attributed to formwork failure. Formwork failures result from faulty formwork structural design, inadequate shoring and reshoring, improper construction practices during construction, inadequate bracing, unstable support or mudsills, and insufficient concrete strength to sustain the applied load after construction.

Contractors are generally responsible for stability and safety of concrete formwork. Contractors are guided by several federal, state, and local codes and regulations that regulate formwork safety. Most of these documents provide general guidelines for safety but provide no guarantee against failure. Contractors typically are trying to achieve fast removal of formwork elements without compromising the safety and integrity of structures.

1.3.3   Quality

The quality of the resulting concrete is dictated by the quality of formwork materials and workmanship. Many concrete-related

problems such as discoloration, stains, and dusting are attributed
to concrete formwork. Also, some deformed concrete surfaces are
due to deformed formwork systems caused by repetitive reuse and
inadequate support of formwork.

1.4 AN INTEGRATED CONCRETE/FORMWORK LIFE CYCLE

The purpose of this section is to introduce formwork operation as
an integrated part of the whole building process and to explain
some of the terminology used in concrete and concrete formwork.
The process of providing formwork and concrete is highly integrated. The left circle in Figure 1.2 represents the formwork life
cycle, while the right circle represents the concrete construction
life cycle. The two intersecting points represent the beginning and
the end of the concrete construction life cycle.

   The life cycle of formwork starts with the "choose formwork"
activity. The physical activities in the formwork life cycle are repre-
sented by these steps: (1) fabricate formwork; (2) erect formwork;
and (3) remove formwork. The concrete construction life cycle

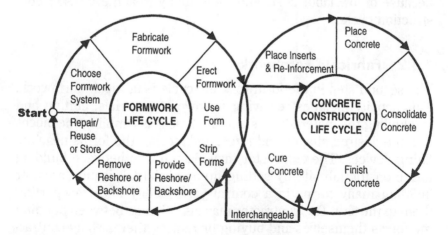

Figure 1.2 Integrated concrete formwork life cycle.

starts after the "fabricate formwork" activity and ends before the "remove formwork" activity. The function of the formwork life cycle is to provide the structure with the specified shape and size, while the function of the concrete construction life cycle is to provide the structure with concrete of specified strength, durability, and surface texture. A brief description of each stage of both the concrete and formwork life cycles is given below.

1.4.1 Choose a Formwork System

The choose formwork system activity includes the process of selecting formwork systems for different structural elements. It also includes the process of selecting accessories, bracing, and a release agent for the selected formwork system. There are several forming systems used in the construction of reinforced concrete structures. For example, formwork systems for concrete slabs can be classified as hand-set or conventional systems and crane-set systems. Conventional systems are still the most common and popular formwork systems. Their popularity stems from their ability to form different shapes and elements. However, conventional formwork usually results in high labor and material cost. Nonconventional or crane-set systems have gained increasing popularity because of low labor costs and their ability to achieve faster construction cycle.

1.4.2 Fabricate Formwork

The second step in the formwork life cycle is fabricate formwork. This activity includes receiving formwork materials, cutting and stockpiling the materials by sizes and types, assembling the pieces into the desired shapes and sizes, and storing the forms near the lifting devices. The contractor may also choose between building forms on the job site by setting up a special fabrication area, or building many forms in a central yard facility and transporting them to the site. The contractor may also choose between building the forms themselves and buying or renting them. Many contractors find that renting forms for specific usage allows them more flexibility in controlling the volume of work they are able to perform.

1.4.3 Erect Formwork, Place Inserts, and Reinforcement

The method and sequence of erecting formwork may vary depending on the availability of lifting equipment and whether reinforcing cages are available. Forms are usually handled manually, by small derrick, or by crane. The erect formwork activity includes the process of lifting, positioning, and aligning the different formwork elements. This activity also includes the process of applying the form release agent or coating that prevents bonding of concrete to forms. The concrete life cycle starts after the erect formwork activity is finished with placing inserts and reinforcement activity. The logical sequencing of erecting formwork and its relation to placing inserts and reinforcement is:

1. Set lines—a template is generally set in place on the floor slab or footing to accurately locate the column floor
2. Erect scaffolding
3. Install column reinforcement
4. Provide forms for column
5. Erect outside forms for walls
6. Install wall reinforcement
7. Erect inside forms for walls
8. Install ties
9. Provide bracing for walls
10. Erect forms for beams
11. Install beam reinforcement
12. Erect forms for slabs
13. Place inserts for mechanical and electrical connections, openings for ducts and conduits, and supporting bars for reinforcement
14. Place secondary and main reinforcement

Figure 1.3 shows inserts and reinforcement installed above the forms.

Figure 1.3 Reinforcement and inserts installed above forms.

A form coating or release agent is often applied to the inside surface of formwork to prevent the concrete from bonding to the formwork elements. Coating can be applied by spraying, brushing, or by a roller. Form coating facilitates the operation of removing the formwork after the concrete has gained enough strength to support itself. Another function of the formwork coating is sealing the surface of the wooden elements which prevent the water in freshly placed concrete from being absorbed by wood. Form release agent should not affect or react with the finished concrete in any way.

1.4.4 Place Concrete

This activity includes mixing, transporting, pumping, and placing of the concrete. The concrete used in most projects is truck-mixed. Concrete is usually transported by belt conveyers for horizontal applications, by buckets for delivery via cranes, by chutes for delivery via gravity to lower levels, and by pumping for horizontal and vertical delivery of concrete.

1.4.5 Consolidate Concrete

Consolidation is the process of compacting or striking the concrete to mold it within the forms, around embedded inserts and reinforcement. It is also done to remove the humps and hollows. Consolidation of concrete is usually performed with hand tools or mechanical vibrators to guarantee a dense structure.

1.4.6 Finish Concrete

This activity includes the process of treating the exposed concrete surfaces to produce the desired appearance, texture, or wearing qualities. Finishing of concrete is usually performed by moving a straight edge back and forth in a sawlike motion across the top of the concrete.

1.4.7 Cure Concrete

The hardening of concrete is a chemical process that requires warmth and moisture. This activity involves curing concrete with water, steam, or any other method to prevent shrinkage and allow the concrete to gain sufficient early strength. Steam curing is used where early strength gain of concrete is important. After the concrete is cured, the rest of the formwork life cycle continues with the strip forms activity. The cure concrete and strip forms activities are interchangeable depending on the type of structural element. For example, columns and walls are cured after stripping of the forms, while slabs and beams are cured before and after the forms are stripped.

1.4.8 Strip Forms

As soon as concrete gains enough strength to eliminate immediate distress or deflection under loads resulting from its own weight and some additional loads, formwork should be stripped to allow other construction activities to start. The operation of removing the forms is called stripping or wrecking the forms. Formwork can either be partially stripped by removing small areas to prevent the slab from deflecting or completely stripped to allow the slab to deflect. As a general rule, formwork supporting members should not be removed before the strength of concrete has reached at least 70 percent of its design value.

1.4.9   Provide Reshores/Backshores

Reshoring and backshoring are the processes of providing temporary vertical support shores for the stripped structural elements which have not yet developed full design strength. They also provide temporary vertical support for the completed structure after the original shoring support has been removed. Reshoring and backshoring are the two methods used to provide the concrete with support until it reaches its full design strength.

Reshores are shores placed snugly under a stripped concrete slab or structural member after the original forms and shores have been removed from a *large* area. In reshoring, the concrete slab is allowed to deflect and, thus, formwork can be removed from a large area. This can help reduce stripping costs, which is the main advantage of reshores.

Backshores are shores placed snugly under a stripped concrete slab or structural member after the original forms and shores have been removed from a *small* area. In backshoring, formwork is removed from a small area of slab and then backshores are provided. Concrete slabs or other structural elements are not allowed to deflect, and as a result, stripping can be accomplished at an earlier concrete curing age.

1.4.10   Remove Reshores or Backshores

Reshores and backshores can be removed after the supported slab or member has attained sufficient strength to support all loads transferred to it. Removal of reshores or backshores must be carried out with care to avoid subjecting the structure to impact loads.

1.4.11 Repair and/or Reuse Formwork

Reuse of concrete formwork is a key for economic formwork construction. After only five reuses, formwork materials costs drop to 40 percent of the initial cost. Formwork elements must be handled with care and should not be dropped. After repairing, cleaning, and oiling, the used formwork elements should either be stockpiled for future use or reused in other areas.

Before reusing formwork elements, they should be inspected for damage. Defects on the inside face must be repaired or removed; otherwise they will reflect on the finished surface of the concrete to show the same defect.

1.5 FORMWORK MATERIALS

Materials used for the construction of concrete formwork range from traditional materials such as wood, steel, aluminum, and plywood to nontraditional materials such as fiberglass. Wood products are the most widely used material for formwork. The objective of this section is to introduce wood as an important material for formwork.

1.5.1 Wood

Wood is widely used for many construction applications including concrete formwork. Wood is harvested from trees and is classified as hardwood and softwood. *Hardwood* comes from trees that have broad leaves such as oaks, maples, and basswood. *Softwood* comes from trees that have needlelike leaves such as pines, cedars, and firs. Softwoods are most commonly used in construction of formwork. It should be noted that the names "hardwood" and "softwood" give no indication of the hardness or the softness of the wood.

Nominal Size

Commercial lumber is sold as boards and planks by dimension sizes. However, the dimensions do not match the actual lumber sizes. For example, a 2 × 4 in. (50.8 × 101.6 mm) pine board is cut to the full 2 × 4 in. (50.8 × 101.6 mm) at the sawmill. This is called the nominal dimension. The nominal dimensions are reduced because of shrinkage and surfacing in both width and thickness. For example, the actual dimensions of a nominal 2 × 4 in. (50.8 × 101.6 mm) are $1^{9}/_{16}$ × $3^{9}/_{16}$ in. (39.7 × 90.5 mm). Lumber that is not surfaced is referred to as rough-sawn. Most lumber for construction is surfaced (dressed) to a standard net size which is less than the nominal (name) size. Surfaced lumber is lumber that has been smoothed or sanded on one side (S1S), two sides (S2S), one edge (S1E), two edges (S2E), or on combinations of sides and edges (S1S1E, S2S1E, S1S2E, or S4S).

Board Measure

Lumber is commonly sold by the foot *board measure*. One board foot is a piece of lumber of 1 in. (25.4 mm) wide, 12 in. (304.8 mm) thick, and 12 in. (304.8 mm) long or its equivalent. The size used in determining board measure is nominal dimension. As a general rule the following formula is used to calculate the foot board measure for lumber:

$$\text{Board feet measure} = \frac{1}{12} (\text{thickness} \times \text{width} \times \text{length})$$

All dimensions should be given in inches.

To find out how much a piece of lumber would cost we have to determine the board feet first and multiply this times the cost. For example, if the cost of pine lumber is \$400 per 1,000 board feet, then the cost of a 2 in. × 6 in. × 10 ft (50.8 × 152.4 × 3048.0 mm) piece would be:

$$\frac{1}{12} \ (2 \text{ in.} \times 6 \text{ in.} \times 10 \text{ ft}) \times \frac{400}{1000} = \$4$$

Commercial Softwood Lumber

There are two major categories for commercial softwood lumber: construction lumber and lumber for remanufacture. Construction lumber is normally used in construction at the same size as it was graded. Grading for construction lumber is normally decided at the sawmill. No further grading occurs once the piece leaves the sawmill. On the other hand, lumber for remanufacture normally undergoes a number of additional manufacturing processes and reaches the consumer in a significantly different form.

Grading of Lumber

Lumber is graded visually and by the machine stress-rated system (MSR). The majority of lumber is graded visually by experienced inspectors who are familiar with lumber grading rules. Lumber grading rules establish limits on the size and characteristics of knots, the number of shakes or splits, and the slope of the grain. Visual evaluation also takes into account any imperfections caused by manufacturing such as torn grain or chip marks. Figure 1.4a shows a typical grade for visually graded lumber.

   In machine stress rating, lumber is evaluated by mechanical stress-rating equipment. Lumber is fed through a machine that subjects each piece of wood to a nondestructive test that measures the modulus of elasticity and bending stress. The machine automatically takes into account size and characteristics of knots, slope of the grain, density, and moisture contents. The machine automatically stamps lumber as *machine rated* and indicates the values for

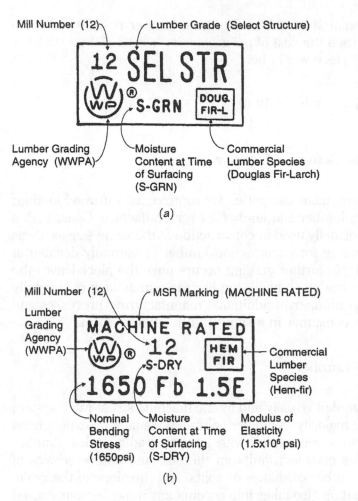

Mill Number (12)   Lumber Grade (Select Structure)

Lumber Grading   Moisture   Commercial
Agency (WWPA)   Content at Time   Lumber Species
                      of Surfacing   (Douglas Fir-Larch)
                      (S-GRN)

(a)

Mill Number (12)   MSR Marking (MACHINE RATED)

Lumber
Grading
Agency
(WWPA)   Commercial
                                        Lumber
                                        Species
                                        (Hem-fir)

Nominal   Moisture   Modulus of
Bending   Content at Time   Elasticity
Stress   of Surfacing   (1.5×10^6 psi)
(1650psi)   (S-DRY)

(b)

Figure 1.4   Gradings of lumber: (a) visually graded; (b) machine rated.

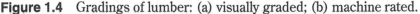

fiber stress in bending and the corresponding modulus of elasticity. Figure 1.4b shows a typical grade for machine-rated (MSR) lumber.

Another method of grading is to grade lumber according to its use and appearance. Lumber is classified as Select, Finish, or Common. Select and Finish are used when fine appearance is required. Common grades are more suitable for general construction.

Size Classification

There are three main size categories for lumber:

1. Boards: Lumber that nominally is less than 2 in. (50.8 mm) thick and 2 in. (50.8 mm) or more wide. *Boards' thicknesses* refers to the smallest cross-section dimension of lumber and the term *width* refers to the largest dimension. Boards less than 6 in. (152.4 mm) wide are classified as strips. Boards are used for sheathing, roofing, siding, and paneling.
2. Dimension lumber: Lumber with a nominal thickness of 2-5 in. (50.8–127 mm) and a nominal width of 2 in. (50.8 mm) or more. Dimension lumber ranges in size from 2 × 2 in. (50.8 × 50.8 mm) to 4 − 16 in. (101.6 × 406.4 mm). Dimension lumber is used for general construction where appearance is not a factor, such as studding, blocking, and bracing.
3. Timber: Lumber that is nominally 5 or more inches (127 mm) in the smallest dimension. Timber is used for columns, posts, beams, stringers, struts, caps, sills, and girders.

Lumber is also grouped according to size and intended use into several categories.

1. Light framing: 2-4 in. (50.8–101.6 mm) thick, 2-4 in. (50.8–101.6 mm) wide. Typical grades are Construction Standard, Utility, and Economy and are widely used for general framing purposes. Lumber under this category is of fine appearance but is graded primarily for strength and serviceability. Utility lumber is used where a combination of high strength and economical construction is

desired. An example would be for general framing, blocking, bracing, and rafters.

2. Studs: 2–4 in. (50.8–101.6 mm) thick, 2–6 in (50.8–152.4 mm) wide, 10 ft (3048.0 mm) long and shorter. Primary use is for walls, whether they are load-bearing or nonload-bearing walls.

3. Structural light framing: 2–4 in. (50.8–101.6 mm) thick and 2–4 in. (50.8–101.6 mm) wide. Typical grades are Select Structural No. 1, No. 2, and No. 3. This is intended primarily for use where high strength, stiffness, and fine appearance are desired. An example of a use would be in trusses.

4. Appearance framing: 2–4 in (50.8–101.6 mm) thick, 2 in. (50.8 mm) and wider. For use in general housing and light construction where knots are permitted but high strength and fine appearance are desired.

5. Structural joists and planks: 2–4 in. (50.8–101.6 mm) thick and 5 in. (127.0 mm) or more wide. Typical grades are Select Structural No. 1, No. 2, and No. 3. Intended primarily for use where high strength, stiffness, and fine appearance are required.

Mechanical Properties of Lumber

Basic understanding of mechanical properties of lumber is necessary for concrete formwork design. Wood is different from any other structural material in that allowable stresses of wood are different according to the orientation of the wood. The intent of the following section is to provide a brief introduction to the mechanical properties of lumber.

Bending Stresses

Figure 1.5 shows a simply supported wood beam with a concentrated load applied at the midpoint. This process results in bend-

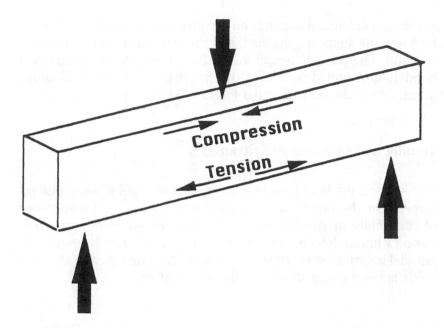

Figure 1.5 Bending stresses.

ing. The lumber is stressed internally to resist the external loads. Bending in a member causes tension forces in the extreme fibers along the face farthest from the load and causes compression in the fiber along the side closest to the applied load. The maximum stress induced in the fibers, which occurs at the edges, is referred to as the "extreme fiber stress in bending." This stress is highly dependent on the parallel-to-grain strength of the wood in both tension and compression. The allowable bending stresses are based on a clear specimen having no defects. Allowable bending stresses are then factored to account for defects.

Modulus of Elasticity (MOE)

Modulus of elasticity is a measure of stiffness. This factor (MOE) is a relationship between the amount of deflection in the member and the value of load applied that causes the deflection. The amount of deflection depends on the size of the member, the span between the supports, the load, and the particular member specie of wood. The parallel-to-grain MOE (i.e., the stiffness when wood is pushed or pulled parallel to the wood grain) is about 30 times greater than the perpendicular-to-grain MOE.

Tensile and Compressive Strengths

Tensile strength is a measure of the ability of wood to resist pulling forces. On the other hand, compressive strength is a measure of the ability of wood to resist pushing forces. For clear wood (wood without defects), the tensile and compressive strengths for parallel-to-grain loads are approximately 10 times greater than for loads applied perpendicular to the wood grain.

Plywood

Plywood is used as sheathing that contacts concrete for job-built forms and prefabricated form panels. Since plywood comes in large sizes, it saves forming time. Plywood is made by gluing together thin layers of wood, called veneer, under intense heat and pressure. Most plywood panels are made of softwood. Many species of trees are used to make plywood, such as Douglas fir and Southern pine. The grain of each ply is laid at a right angle to the adjacent pieces. This process gives plywood extra strength and reduces shrinkage and swelling.

Plywood is commonly sold in large sheets 4 × 8 ft (1.22 × 2.44 m). These large panels reduce erection and stripping costs and produce fewer joints on the finished concrete. Plywood is available in varieties of thicknesses that identify it for sale. For example, plywood called "½ in. (12.7 mm) plywood" is ½ in. (12.7 mm) thick. In contrast to the situation with lumber, actual and nominal thickness for plywood is the same; 1 in. (25.4 mm) plywood is 1 full inch (25.4 mm) thick.

Plyform

Plyform is a plywood product specially made for concrete formwork. Plyforms are available in class I and class II, where class I
is stronger than class II. Other plyform that is commonly used for
formwork includes B-B plyform, high-density overlaid (HDO), and
structural 1 plyform.

Plywood and plyform have particular orientations that affect
their strength. A weak position can be achieved when the grain
(face grain) is parallel to the span of support. A stronger orientation is seen when the grain (face grain) is perpendicular to the
span of support (Figure 1.6).

1.5.2   Steel

The major advantages of steel sections in formwork are the ability
of steel to form longer spans and its indefinite potential for reuse
when handled with reasonable care. Steel sections are used in the
fabrication of different formwork components, namely: (1) steel
panel forms, (2) horizontal and vertical shores, (3) steel pan and
dome components used for joist and waffle slabs, and (4) steel
pipes for formwork bracing. Other heavy forms and formwork are
also made of steel, such as bridge formwork. Steel is used for formwork when other materials are impossible to use because of their
low strength. Steel forms are typically patented, and allowable
loads are generally published by the manufacturers. Figure 1.7 il-
lustrates the use of steel as formwork materials

1.5.3   Aluminum

Aluminum has become an increasingly popular material for many
formwork applications such as lightweight panels, joists, horizontal and vertical shoring, and aluminum trusses for flying forms.
The popularity of aluminum stems from its light weight which reduces handling costs and offsets its higher initial material cost.
When compared to steel panels, aluminum panels used for ganged
forms weigh approximately 50 percent less. The major problem

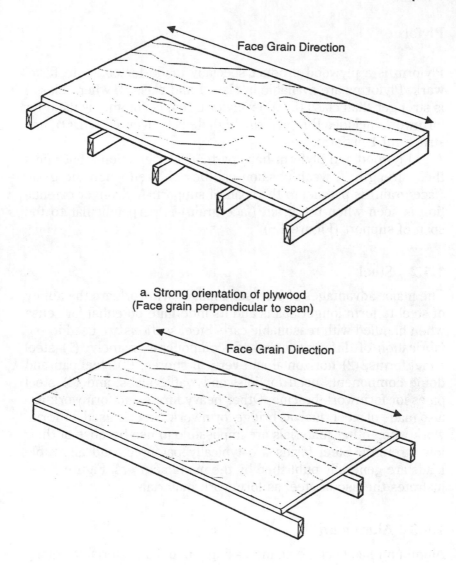

a. Strong orientation of plywood
(Face grain perpendicular to span)

b. Weak orientation of plywood
(Face grain perpendicular to span)

Figure 1.6 Plywood orientation.

Figure 1.7   Steel formwork.

with aluminum forms is corrosion: Pure aluminum is attacked chemically by wet concrete. Aluminum alloys have proven to be very successful in resisting corrosion. Figure 1.8 shows aluminum panels used for wall forms.

1.5.4   Glass-Reinforced Plastic

In recent years, forms fabricated from glass-reinforced plastic have found increasing use because of their strength, light weight, and high number of reuses. Glass-reinforced plastic also produces high-quality concrete finishes. Glass-reinforced plastic forms are very flexible and can form complex or nonstandard shapes with little capital investment.

To fabricate glass-reinforced plastic forms, models of plaster, wood, or steel are prepared to the exact desired dimensions. The model is then waxed, polished, and sprayed with a parting agent to prevent sticking of the resin to the master pattern. Glass mat is then fitted over the model and thoroughly saturated with a brush

Figure 1.8 Aluminum formwork.

coat of polyester resin. When the resin has set and the heat dissi-
pated, another layer of glass mat and polyester resin is added, and
this process is repeated until the desired thickness of the fiberglass sheet is achieved.

Another method to build glass-reinforced plastic forms is
through the use of a spray gun to apply the resin to chopped
strands of fiberglass, which are used as the reinforcing material.

To increase the number of potential reuses with any of the
methods of fabrication mentioned, an extra thickness of resin is
molded into the contact surface or additional stiffening and supports are added by means of built-up ribs, wood struts, steel rods,
or aluminum tubing.

The two major problems associated with glass-reinforced
plastic forms are attack by alkalies in the concrete and form expan-
sion because of exposure to hot sun or heat from hydration of
cement.

2
Horizontal Formwork Systems: Hand-Set Systems

2

Horizontal Formwork Systems: Hand-Set Systems

2.1 HORIZONTAL FORMWORK SYSTEMS CLASSIFICATION

Horizontal formwork systems are used to temporarily support horizontal concrete work such as concrete slabs. There are seven horizontal forming systems that can be used to support different slab types. They are: (1) conventional wood system (stick form), (2) conventional metal (aluminum) system (improved stick form), (3) flying formwork system, (4) column-mounted shoring system, (5) tunnel forming system, (6) joist-slab forming system, and (7) dome forming system. Joist-slab and dome forms are steel or fiberglass pans usually placed above the plywood sheathing and thus can be used with any of the first five horizontal formwork systems. As a result, they will not be considered in this book as separate systems.

Formwork systems for horizontal concrete work can be also classified into two main categories: hand-set systems and crane-set systems. Conventional wood systems and conventional metal systems are classified as hand-set systems. In hand-set systems, different formwork elements can be handled by one or two laborers. Flying formwork systems, column-mounted shoring systems, and tunnel formwork are classified under crane-set systems. In crane-set systems, adequate crane services must be available to handle formwork components.

2.2   CONVENTIONAL WOOD FORMWORK SYSTEM

The conventional wood system is sometimes referred to as the stick form or hand-over-hand method. Conventional wood system includes formwork for slabs, beams, and foundations. The system is generally built of lumber or a combination of lumber and plywood. Formwork pieces are made and erected in situ. For stripping, conventional wood systems are stripped piece by piece, then cleaned, and may be reused a few times.

2.2.1   Formwork for Concrete Slabs

Conventional wood systems for horizontal concrete work are made of plywood or lumber sheathing for decking. As it will be discussed in Chapter 3, the thickness of plywood or lumber is determined by structural analysis and is a function of the applied loads, type of wood or plywood, and the spacing between sheathing supporting elements. Plywood is preferred over lumber sheathing because it provides a smooth concrete surface that requires minimum finishing effort. The use of plywood for decking is also productive because of its large panel size (4 × 8 in.) (1.22 × 2.44 m).

Sheathing is supported by horizontal members called joists or runners. Joists are made from dimension lumber spaced at constant intervals that are a function of applied loads and the type of lumber. It is a recommended practice to round down the calculated joist spacing to the lower modular value.

Joists are supported by another set of horizontal members perpendicular to the joists, called stringers. The stringers are supported by vertical members called shores. In all-wood conventional formwork systems, shores are made of dimension lumber that have square cross sections [i.e., 4 × 4 in. (101.6 × 101.6 mm) or 6 × 6 in. (152.4 × 152.4 mm)]. Shores are rested on heavy timbers, called mudsills, to transfer the vertical loads to the ground. In the case where a slab-on-grade exists, shores are rested directly on them. Figure 2.1 shows a typical all-wood conventional formwork system for concrete slabs.

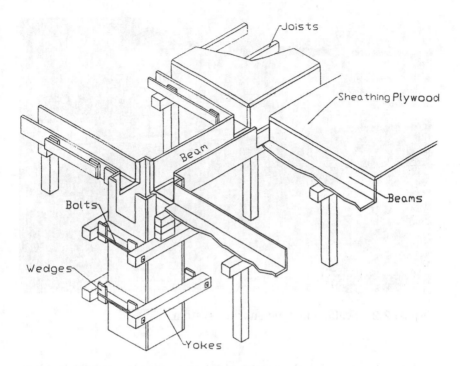

Figure 2.1 All-wood conventional formwork system.

Vertical timber shores can be replaced by the scaffold type, which has been proven to be more efficient because of its high number of reuses and its height, which means that no splicing is typically required. The scaffold-type shoring system consists of two vertical steel posts with horizontal pipe between them at regular intervals. Adjustable screw jacks are fitted into the steel posts at both ends. The top jacks are fitted into steel caps called tee-heads. The bottom jacks are fastened into rectangular steel plates. Adjacent vertical steel posts are braced together by steel X braces. Figure 2.2 shows a typical scaffold-type shoring system.

2.2.2 Formwork for Concrete Beams

Formwork for beams consists of a bottom and two sides in addition to their supporting elements. The bottom is typically made of ply-

Figure 2.2   Scaffolding-type shoring system.

wood or lumber sheathing with thickness of 0.75 in. (19.0 mm) or
1 in. (25.4 mm). The bottom is supported by and fastened to horizontal joists. Beam sides are also made of plywood or lumber
sheathing.

Once the bottom of the beam form is constructed and leveled,
one side of the beam is erected first with holes drilled into it for
installing the tie rods. Tie rods are steel rods that hold the two
sides of the beam together. After the first side of the beam form
is erected, the reinforcement is placed inside the beam and then
the other side of the beam is erected. Tie rods are then inserted
into all holes and the walers on both sides of the beam. The tie
rods' function is to resist the horizontal pressure resulting from
the freshly placed concrete and thus keep the sides of the beams
in their proper location. Tie rods are fastened to the sides of the
beam and also to vertical walers and clamps. To further support
the two sides of the beam and hold them together, additional temporary spreaders are fastened at the top of the beam sides at regu-

lar distances. Temporary spreaders may be made from wood or steel.

2.2.3   Formwork for Foundation

Formwork for continuous or isolated footing is usually made of wood boards (planks) or plywood supported by vertical stakes driven into the ground. The top of the vertical stake is supported by a diagonal brace driven into the ground. Bracing may be replaced by the piling up of dirt to support the sides of the form. The correct distance between the planks is kept by crossing spreaders. On small footings, steel straps are used to replace the spreaders. It is common practice to construct the forms higher than necessary and place the straps in the inner sides of the form at a height equal to the concrete level. Figure 2.3 shows typical formwork for isolated footing.

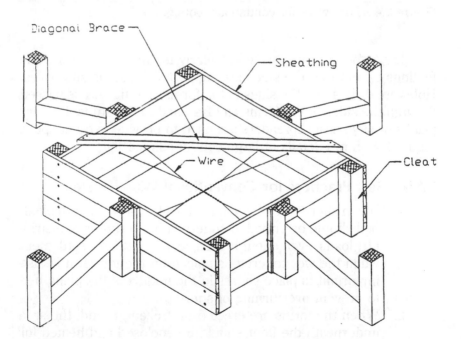

Figure 2.3   Formwork for isolated footing.

Figure 2.4   Formwork for continuous footing.

Large footings are formed similarly to small and continuous footings except that the sides are supported by studs and wales. Holes are drilled into the sides of the forms, and tie wire is passed through the sides of the forms and fastened to the studs. Lumber planks or steel strap spreaders are used to provide extra support. Figure 2.4 shows details of large continuous footing.

2.2.4   Best Practices for Conventional Wood System

1. When shores are rested on soft soil, a large enough plank bed should be provided underneath the shores to distribute loads over enough area to prevent any settlement when the wet concrete is placed on the forms. It is also important to place shores in the middle of the plank bed to prevent overturning of shores.
2. When the forms are erected on frozen ground, the area underneath the floor should be enclosed and heated for enough time before the placing of concrete to ensure the

removal of frost and to provide a stable foundation for the forms.

3. Beams, girders, and sometimes long slab forms should be given a slight camber to reduce any visible deflection after the placing of concrete.

4. It is important to leave one side of the column form open to clean out shavings or rubbish. The open side is immediately closed before concrete is placed. In deep, narrow forms, holes should be provided at the bottom for cleaning and inspection.

5. On less important work, it is normal practice to wet the forms immediately before placing concrete. On large jobs where forms are to be reused several times, form surfaces should be oiled or coated with form coating. Oiling or coating should be done before the reinforcement is placed to prevent greasing the steel, which reduces or eliminates the bonding between the steel and concrete. Coating should not be so thick as to stain the concrete surface.

2.2.5   Limitations of Conventional Wood System

There are three major problems with use of all-wood conventional formwork systems.

1. *High labor costs.* The conventional formwork system is a labor-intensive system. Labor costs range from 30 to 40 percent of the total cost of concrete slabs.

2. *High waste.* Erecting and dismantling conventional formwork is conducted piece by piece. This causes breaking of edges and deformation of wood. It is estimated that 5 percent waste is generated from a single use of formwork.

3. *Limited number of reuses.* Number of reuses is the key to an economical formwork construction. Typically, conventional formwork is limited to five to six reuses. A limited number of reuses forces the contractor to use several sets

of formwork; this adds to the expense of formwork construction.

4. *Higher quality of labor force and supervision.* Conventional formwork systems work best with a high-quality labor force and adequate supervision. In areas with an unskilled or semiskilled labor force and minimal supervision, more sophisticated formwork systems are more appropriate.

5. *Limited spans.* Since dimension wood is low strength compared to that of aluminum and steel sections, it has limited use in applications where long spans are desired.

2.2.6 Advantages of Conventional Wood System

Despite the limitations of the conventional wood system, it has a few distinct advantages.

1. *Flexibility.* Because the system is built piece by piece, it is virtually capable of forming any concrete shape. A complicated architectural design can be formed only by this system.

2. *Economy.* This system is not economical in terms of labor productivity and material waste. However, the system may be economical for small projects with limited potential reuses. The system has the advantage of low makeup cost or initial cost. Also, for restricted site conditions, where storage areas are not available and the use of cranes is difficult, the conventional wood system might be the only feasible alternative. (It is interesting to note that a skyscraper built in the late 1980s in New York City used the conventional wood system because of restricted site conditions.)

3. *Availability.* Wood is a construction material that is available virtually anywhere. In areas where formwork suppliers are not available, a conventional wood system may be the only feasible alternative. Availability and low labor cost are the two main reasons behind the popularity of

the conventional wood system in the developed coun-
tries.

2.3   CONVENTIONAL METAL SYSTEMS

In the conventional metal system, joists and stringers are made of
aluminum or steel supported by scaffold-type aluminum or steel
shoring. In today's construction practices, joists and stringers are
made of aluminum and are supported by a scaffold-type movable
shoring system. In this book, the term *conventional aluminum system* is used to describe the latter system.

2.3.1   Types of Conventional Metal Systems

The conventional aluminum system, described above, is widely
used and is selected as an example of conventional metal systems
for purposes of comparison with other horizontal formwork systems. Other conventional metal systems include different combinations of wood, aluminum, and steel for joists and stringers. In
all the systems listed below, a movable steel-scaffolding system or
single steel post is used for shoring.

1.   Plywood sheathing for decking supported by dimen-
     sional wood or laminated wood for joists and standard
     steel sections for stringers.
2.   Plywood sheathing for decking supported by standard
     steel sections for joists and stringers. A movable alumi-
     num or steel scaffolding system is used for shoring.

In these two systems, steel joists and stringers have the advantage of supporting greater spans, resulting in fewer vertical
shores and fewer joists and stringers. The main problem with using steel as joists and stringers for forming concrete slabs is their
heavy weight, which makes it difficult for one person to handle.

A standard steel W-section is used because its wide flange
makes it easy to connect stringers with shore legs. It should be
noted that stringers have to be well secured to the shore to prevent

stringers from being overturned. A special accessory is used to connect the steel beam to the shore head.

2.3.2   Conventional Aluminum System Description

The conventional aluminum system is sometimes referred to as "an improved stick system." The system is a hand-set system that consists of a deck surface made of plywood or wood supported by aluminum "nailer type" joists and stringers as shown in Figure 2.5. The same type of deck forms can be made up of large panels tied or ganged together and supported by steel scaffold-type shoring. Aluminum panel models range in length from 2 to 8 ft (0.61 to 2.44 m), and in width from 2 to 36 in. (50.8 to 914.4 mm).

Plywood Sheathing

Plywood and plyform can be used as sheathing. As indicated in Chapter 1, plywood is available in two types, interior and exterior.

Figure 2.5   Conventional aluminum system.

Exterior plywood is used for sheathing because it is made of water-proof glue that resists absorption of concrete-mix water. The thickness of the plywood is a design function; however, 0.75-in.-thick (19.0 mm) plywood is widely used for concrete slabs.

Extruded Aluminum Joist

The first component of the conventional metal system is the aluminum joist. The extruded aluminum joist takes the shape of a modified I beam with a formed channel in the top flange in which a wood nailer strip 2 × 3 in. (50.8 × 76.2 mm) is inserted; the plywood deck is then nailed to the nailer strip. Figure 2.6 shows two different shapes of extruded aluminum joist.

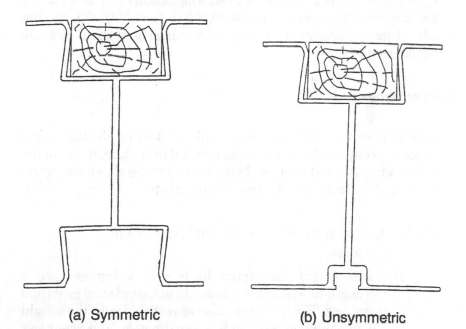

(a) Symmetric   (b) Unsymmetric

Figure 2.6 Nailer-type joists: (a) symmetric; (b) Unsymmetric.

Aluminum Beams (Stringers)

The purpose of stringers is to transfer the loads of the surface panels to the scaffold. Extruded aluminum joists can be also used as stringers unless the loading is too excessive; in fact, it is good practice to avoid using a mixture of different beam types. Aluminum beams are commercially available for lengths ranging from 4 to 30 ft (1.22 to 9.14 m) in 2-ft (0.6 1-m) increments.

Aluminum Scaffold Shoring

The aluminum scaffold shoring system has been available for several years as a substitute for the steel scaffold shoring system. The system consists of several frames connected together by cross bracing. Aluminum shoring is lighter and has load-carrying capacity equal to or greater than steel shoring. Load-carrying capacity for aluminum shoring can reach 36,000 lb (160 kN). Despite the advantages of aluminum shoring, steel shoring systems are still widely used.

Post Shore

A post shore is a single member made of steel or aluminum and support stringers. Post shores can be used to replace, or in combination with, scaffold shoring. Post shores can be also used for reshoring after stripping of formwork elements.

2.3.3   Advantages of Conventional Aluminum Systems

1. *Light weight.* Aluminum joists and stringers have a strength-to-weight ratio better than that of steel joists and stringers. For the same value of vertical loads, the weight of an aluminum section is approximately 50 percent less than that of the corresponding steel, ranging from 3 to 6

lb/ft (4.5 to 8.9 kg/m) including the wood nailing strip. Also, the light weight of the aluminum improves labor productivity and reduces crane time hook in comparison to steel sections.

2. *Economy.* Many contractors have reported labor cost saving of between 20 to 30 percent for both fabricating and stripping aluminum formwork compared to conventional wood form.

3. *High reuse.* Aluminum sections have higher reuse value than wood sections. Also, the amount of waste is minimal compared to that of wood sections.

2.3.4 Limitations of Conventional Aluminum Systems

Despite the many improvements of this system over the conventional wood system, it is still considered a labor-intensive system. Fabrication, erection, and stripping are done piece by piece, resulting in low labor productivity and high waste of plywood. Generally, aluminum sections and aluminum or steel scaffold are durable unless sections are exposed to rough handling on site, especially during stripping of formwork.

Another problem associated with the general use of aluminum with concrete is the chemical reaction between aluminum and spilled concrete.

2.4 SPECIAL HORIZONTAL FORMWORK SYSTEM

Some special structural slabs such as the one-way and the two-way joist slabs require the use of flange-type pan forms (Figure 2.7). Pans are usually nailed to supporting joists (soffits) or to sheathing. Pans nailed to sheathing are preferred because they make layout easier, the work area more efficient, and they are safer for those working on and below the deck. Pans are typically placed overlapping from 1 to 5 in. (25.4 to 127 mm). Pans are usually

Figure 2.7 Flange-type pan forms.

stripped manually or by using compressed air applied to an adapter at the center of the dome.

2.4.1 Joist-Slab Forming Systems

A one-way joist slab is a monolithic combination of regularly spaced joists arranged in one direction and a thin slab cast in place to form an integral unit with the beams and columns (Figure 2.8a). One-way joist slabs have frequently been formed with standard steel pans. Table 2.1 shows the dimensions of the standard-form pans and the special fillers for one-way joist construction. Any spacing between pans which exceeds 30 in. (762 mm) is referred to as a wide-modular or skip-joist system.

2.4.2 Dome Forming Systems

Standard size domes are usually used for waffle slab construction. They are based on 2-ft (0.61-m), 3-ft (0.91-m), 4-ft (1.22-m), and

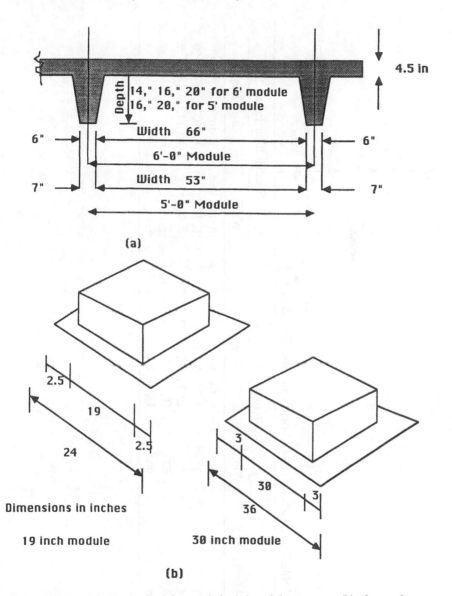

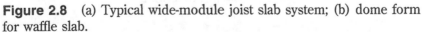

Figure 2.8   (a) Typical wide-module joist slab system; (b) dome form for waffle slab.

Table 2.1 Dimensions of Forms for One-Way Joist Slab

Module (ft)	Standard forms (in.)		Special filler forms (in.)	
	Width	Depth	Width	Depth
2	20	8, 10, 12	10, 15	8, 10, 12
3	30	8, 10, 12, 14, 16, 20	10, 15, 20	8, 10, 12, 14, 16, 20
4	40	12, 14, 16, 18, 20, 22, 24	20, 30	12, 14, 16, 18, 20, 22, 24
5	53	16, 20	—	—
6	66	14, 16, 20	—	—

5-ft (1.52-m) modules. The 2-ft (0.61-m) size modules utilize 19 × 19 in. (482.6 × 482.6 mm) domes, with 5-in. (127.0-mm) ribs between them, and the 3-ft. (0.91-m) size modules can be formed with 30 × 30 in. (762 × 762 mm) domes and 6-in. (152.4-mm) ribs. Figure 2.8b shows the two standard modules that are used for waffle slab construction.

3
Slab Form Design

Concrete forms are engineered structures that are required to support loads composed of fresh concrete, construction materials, equipment, workers, impact of various kinds, and sometimes wind. The forms must support all the applied loads without collapse or excessive deflection. ACI Committee Report 347-1994 defines those applied loads and gives a number of guidelines for safety and serviceability. Based on these guidelines, a number of design tables have been developed for the design of concrete formwork. These tables are useful design tools. However, they do not take into consideration stress modification factors that are provided by the National Design Specification for Wood Construction, NDS 1991. This chapter presents a design procedure for all-wood concrete slab forms based on NDS 1991 and Plywood Design Specification 1997.

The objective of the formwork design is to determine the safe spacing for each slab form component (sheathing, joists, stringers, and shores), and ensure that each component has adequate strength to resist the applied pressure without exceeding predetermined allowable deflection.

3.1 PROPERTIES OF FORM MATERIALS

The following sections provide an overview of some important properties of structural sections that are used in formwork design. Readers familiar with these expressions should start with Section 3.3.

3.2   PROPERTIES OF AREA

Certain mathematical expressions of the properties of sections are
used in design calculations for various design shapes and loading
conditions. These properties include the moment of inertia, cross
sectional area, neutral axis, section modulus, and radius of gyration of the design shape in question. These properties are described below.

1.   *Moment of inertia.* The moment of inertia I of the cross
     section is defined as the sum of the products of the differ-
     ential areas into which the section may be divided, multi-
     plied by the squares of their distances from the neutral
     axis of the section (Figure 3.1).
        If the section is subjected to a bending moment about
     the X-X axis of the cross section, the moment of inertia
     about X-X is denoted by I_{xx},

$$I_{xx} = \sum_{i=1}^{n} A_i Y_i^2$$

     where

        n = total number of differential areas
        A_i = area of element i
        Y_i = distance between element i and X-X axis

     If the member is subjected to a bending moment about
     axis Y-Y of the cross section, we denote the moment of
     inertia associated with it as I_{yy},

$$I_{yy} = \sum_{j=1}^{m} A_j X_j^2$$

     where

        m = total number of elementary areas
        A_j = area of element j
        X_j = distance between element j and Y-Y axis

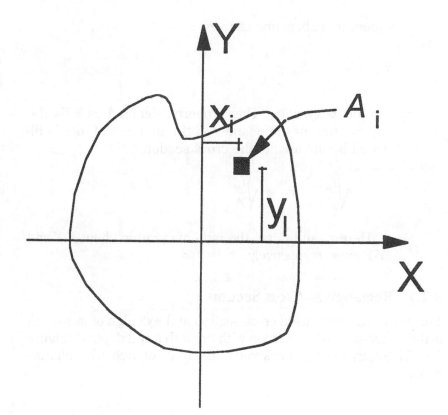

Figure 3.1 Moment-of-inertia calculation.

2. *Cross sectional area.* This is the area of a section taken through the member, perpendicular to its longitudinal axis.

3. *Neutral axis.* The neutral axis is a line through the cross section of the member along which the fibers sustain neither tension nor compression when subjected to a loading.

4. *Section modulus.* Denoted as S, this is the moment of inertia divided by the distance between the neutral axis and the extreme fibers (maximum stressed fibers) of the cross section.

    If c is the distance from the neutral axis to the extreme

fibers in inches, one can write:

$$S_{xx} = \frac{I_{xx}}{c} \qquad S_{yy} = \frac{I_{yy}}{c}$$

5. *Radius of gyration*. This property, denoted as r, is the square root of the quantity of the moment of inertia divided by the area of the cross section.

$$r_{xx} = \sqrt{\frac{I_{xx}}{A}} \qquad r_{yy} = \sqrt{\frac{I_{yy}}{A}}$$

Here r_{xx} and r_{yy} are the radii of gyration about X-X and Y-Y axes, respectively.

3.2.1 Rectangular Cross Section

The most commonly used cross section in the design of formwork is the rectangular cross section with breadth b and depth d (Figure 3.2). These are usually measured in the units of inches or millimeters.

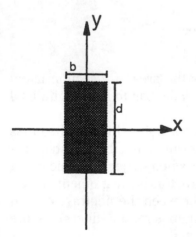

Figure 3.2 Rectangular cross section.

For rectangular cross section, the formulas discussed in the previous section take the forms:

Moments of intertia: $I_{xx} = \dfrac{bd^3}{12}$, in.^{4} or mm^4

$$I_{yy} = \dfrac{db^3}{12}, \text{in.}^4 \text{or mm}^4$$

Radii of gyration: $r_{xx} = \sqrt{\dfrac{I_{xx}}{A}} = \dfrac{d}{\sqrt{12}}$, in. or mm

$$r_{yy} = \sqrt{\dfrac{I_{yy}}{A}} = \dfrac{b}{\sqrt{12}}, \text{in. or mm}$$

Section modules:

$$S_{xx} = \frac{I_{xx}}{c} = \frac{bd^2}{6}, \text{in.}^3 \text{or mm}^3 \left(\text{here } c = \frac{d}{2}\right)$$

$$S_{yy} = \frac{I_{yy}}{c} = \frac{db^2}{6}, \text{in.}^3 \text{or mm}^3 \left(\text{here } c = \frac{b}{2}\right)$$

The section properties for selected standard sizes of board, dimension lumber, and timbers are given in Table 3.1. The values given in this table can be used to calculate the properties given above. Table 3.2 provides section properties of standard dressed (S4S) sawn lumber.

3.3   PROPERTIES OF SAWN LUMBER

3.3.1   Classification of Sawn Lumber

Structural Sawn Lumber size classification was discussed in Chapter 1 and is summarized below.

1. Dimension: 2 in. < thickness < 4 in. and width > 2 in.
2. Beams and stringers: thickness > 5 in. and width > thickness + 2 in.

Table 3.1   Nominal and Minimum Dressed Sizes of Sawn Lumber

	Thickness (in.)			Face widths (in.)		
		Minimum dressed			Minimum dressed	
Item	Nominal	Dry	Green	Nominal	Dry	Green
Boards	1	3/4	25/32	2	1-1/2	1-9/16
	1-1/4	1	1-1/32	3	2-1/2	2-9/16
	1-1/2	1-1/4	1-9/32	4	3-1/2	3-9/16
				5	4-1/2	4-5/8
				6	5-1/2	5-5/8
				7	6-1/2	6-5/8
				8	7-1/4	7-1/2
				9	8-1/4	8-1/2
				10	9-1/4	9-1/2
				11	10-1/4	10-1/2
				12	11-1/4	11-1/2
				14	13-1/4	13-1/2
				16	15-1/4	15-1/2
Dimension Lumber	2	1-1/2	1-9/16	2	1-1/2	1-9/16
	2-1/2	2	2-1/16	3	2-1/2	2-9/16
	3	2-1/2	2-9/12	4	3-1/2	3-9/16
	3-1/2	3	3-1/16	5	4-1/2	4-5/8
				6	5-1/2	5-5/8
				8	7-1/4	7-1/2
				10	9-1/4	9-1/2
				12	11-1/4	11-1/2
				14	13-1/4	13-1/2
				16	15-1/4	15-1/2
Dimension Lumber	4	3-1/2	3-9/16	2	1-1/2	1-9/16
	4-1/2	4	4-1/16	3	2-1/2	2-9/16
				4	3-1/2	3-9/16
				5	4-1/2	4-5/8
				6	5-1/2	5-5/8
				8	7-1/4	7-1/2
				10	9-1/4	9-1/2
				12	11-1/4	11-1/2
				14	—	13-1/2
				16	—	15-1/2
Timbers	5 and thicker	—	1/2 off	5 and wider	—	1/2 off

From National Design Specification for Wood Construction 1991

Table 3.2 Section Properties of Standard Dressed (S4S) Sawn Lumber

Nominal size b × d (inches × inches)	Standard dressed size (S4S) b × d (inches × inches)	Area of Section A in²	X-X-AXIS Section modulus S_{xx} in³	X-X-AXIS Moment of inertia I_{xx} in⁴	Y-Y-AXIS Section modulus S_{yy} in³	Y-Y-AXIS Moment of inertia I_{yy} in⁴	Approx. weight (lb/ft) 25 lb/ft³	30 lb/ft³	35 lb/ft³	40 lb/ft³	45 lb/ft³	50 lb/ft³
1 × 3	3/4 × 2-1/2	1.875	0.781	0.977	0.234	0.088	0.326	0.391	0.456	0.521	0.586	0.651
1 × 4	3/4 × 3-1/2	2.625	1.531	2.680	0.328	0.123	0.456	0.547	0.638	0.729	0.820	0.911
1 × 6	3/4 × 5-1/2	4.125	3.781	10.40	0.516	0.193	0.716	0.859	1.003	1.146	1.289	1.432
1 × 8	3/4 × 7-1/4	5.438	6.570	23.82	0.680	0.255	0.944	1.133	1.322	1.510	1.699	1.888
1 × 10	3/4 × 9-1/4	6.938	10.70	49.47	0.867	0.325	1.204	1.445	1.686	1.927	2.168	2.409
1 × 12	3/4 × 11-1/4	8.438	15.82	88.99	1.055	0.396	1.465	1.758	2.051	2.344	2.637	2.930
2 × 3	1-1/2 × 2-1/2	3.750	1.563	1.953	0.938	0.703	0.651	0.781	0.911	1.042	1.172	1.302
2 × 4	1-1/2 × 3-1/2	5.250	3.063	5.359	1.313	0.984	0.911	1.094	1.276	1.458	1.641	1.823
2 × 5	1-1/2 × 4-1/2	6.750	5.063	11.39	1.688	1.266	1.172	1.406	1.641	1.875	2.109	2.344
2 × 6	1-1/2 × 5-1/2	8.250	7.563	20.80	2.063	1.547	1.432	1.719	2.005	2.292	2.578	2.865
2 × 8	1-1/2 × 7-1/4	10.88	13.14	47.63	2.719	2.039	1.888	2.266	2.643	3.021	3.398	3.776
2 × 10	1-1/2 × 9-1/4	13.88	21.39	98.93	3.469	2.602	2.409	2.891	3.372	3.854	4.336	4.818
2 × 12	1-1/2 × 11-1/4	16.88	31.64	178.0	4.219	3.164	2.930	3.516	4.102	4.688	5.273	5.859
2 × 14	1-1/2 × 13-1/4	19.88	43.89	290.8	4.969	3.727	3.451	4.141	4.831	5.521	6.211	6.901
3 × 4	2-1/2 × 3-1/2	8.750	5.104	8.932	3.646	4.557	1.519	1.823	2.127	2.431	2.734	3.038
3 × 5	2-1/2 × 4-1/2	11.25	8.438	18.98	4.688	5.859	1.953	2.344	2.734	3.125	3.516	3.906
3 × 6	2-1/2 × 5-1/2	13.75	12.60	34.66	5.729	7.161	2.387	2.865	3.342	3.819	4.297	4.774
3 × 8	2-1/2 × 7-1/4	18.13	21.90	79.39	7.552	9.440	3.147	3.776	4.405	5.035	5.664	6.293
3 × 10	2-1/2 × 9-1/4	23.13	35.65	164.9	9.635	12.04	4.015	4.818	5.621	6.424	7.227	8.030
3 × 12	2-1/2 × 11-1/4	28.13	52.73	296.6	11.72	14.65	4.883	5.859	6.836	7.813	8.789	9.766
3 × 14	2-1/2 × 13-1/4	33.13	73.15	484.6	13.80	17.25	5.751	6.901	8.051	9.201	10.35	11.50
3 × 16	2-1/2 × 15-1/4	38.13	96.90	738.9	15.89	19.86	6.619	7.943	9.266	10.59	11.91	13.24
4 × 4	3-1/2 × 3-1/2	12.25	7.146	12.51	7.146	12.51	2.127	2.552	2.977	3.403	3.828	4.253
4 × 5	3-1/2 × 4-1/2	15.75	11.81	26.58	9.188	16.08	2.734	3.281	3.828	4.375	4.922	5.469
4 × 6	3-1/2 × 5-1/2	19.25	17.65	48.53	11.23	19.65	3.342	4.010	4.679	5.347	6.016	6.684
4 × 8	3-1/2 × 7-1/4	25.38	30.66	111.1	14.80	25.90	4.405	5.286	6.168	7.049	7.930	8.811
4 × 10	3-1/2 × 9-1/4	32.38	49.91	230.8	18.89	33.05	5.621	6.745	7.869	8.993	10.12	11.24
4 × 12	3-1/2 × 11-1/4	39.38	73.83	415.3	22.97	40.20	6.836	8.203	9.570	10.94	12.30	13.67
4 × 14	3-1/2 × 13-1/2	47.25	106.3	717.6	27.56	48.23	8.203	9.844	11.48	13.13	14.77	16.41
4 × 16	3-1/2 × 15-1/2	54.25	140.1	1086.1	31.64	55.38	9.42	11.30	13.19	15.07	16.95	18.84

Table 3.2 Continued

Nominal size $b \times d$	Standard dressed size (S4S) $b \times d$ inches × inches	Area of Section A in²	X-X-AXIS Section modulus S^{xx} in³	X-X-AXIS Moment of inertia I_{xx} in⁴	Y-Y-AXIS Section modulus S_{yy} in³	Y-Y-AXIS Moment of inertia I_{yy} in⁴	Approximate weight in pounds per linear foot (lb/ft) of piece when density of wood equals: 25 lb/ft³	30 lb/ft³	35 lb/ft³	40 lb/ft³	45 lb/ft³	50 lb/ft³
5 × 5	4-1/2 × 4-1/2	20.25	15.19	34.17	15.19	34.17	3.516	4.219	4.922	5.625	6.328	7.031
6 × 6	5-1/2 × 5-1/2	30.25	27.73	76.26	27.73	76.26	5.252	6.302	7.352	8.403	9.453	10.50
6 × 8	5-1/2 × 7-1/2	41.25	51.56	193.4	37.81	104.0	7.161	8.594	10.03	11.46	12.89	14.32
6 × 10	5-1/2 × 9-1/2	52.25	82.73	393.0	47.90	131.7	9.071	10.89	12.70	14.51	16.33	18.14
6 × 12	3-1/2 × 11-1/2	63.25	121.2	697.1	57.98	159.4	10.98	13.18	15.37	17.57	19.77	21.96
6 × 14	5-1/2 × 13-1/2	74.25	167.1	1128	68.06	187.2	12.89	15.47	18.05	20.63	23.20	25.78
6 × 16	5-1/2 × 15-1/2	85.25	220.2	1707	78.15	214.9	14.80	17.76	20.72	23.68	26.64	29.60
6 × 18	5-1/2 × 17-1/2	96.25	280.7	2456	88.23	242.6	16.71	20.05	23.39	26.74	30.08	33.42
6 × 20	5-1/2 × 19-1/2	107.3	348.6	3398	98.31	270.4	18.62	22.34	26.07	29.79	33.52	37.24
6 × 22	5-1/2 × 21-1/2	118.3	423.7	4555	108.4	298.1	20.53	24.64	28.74	32.85	36.95	41.06
6 × 24	5-1/2 × 23-1/2	129.3	506.2	5948	118.5	325.8	22.44	26.93	31.41	35.90	40.39	44.88
8 × 8	7-1/2 × 7-1/2	56.25	70.31	263.7	70.31	263.7	9.766	11.72	13.67	15.63	17.58	19.53
8 × 10	7-1/2 × 9-1/2	71.25	112.8	535.9	89.06	334.0	12.37	14.84	17.32	19.79	22.27	24.74
8 × 12	7-1/2 × 11-1/2	86.25	165.3	950.5	107.8	404.3	14.97	17.97	20.96	23.96	26.95	29.95
8 × 14	7-1/2 × 13-1/2	101.3	227.8	1538	126.6	474.6	17.58	21.09	24.61	28.13	31.64	35.16
8 × 16	7-1/2 × 15-1/2	116.3	300.3	2327	145.3	544.9	20.18	24.22	28.26	32.29	36.33	40.36
8 × 18	7-1/2 × 17-1/2	131.3	382.8	3350	164.1	615.2	22.79	27.34	31.90	36.46	41.02	45.57
8 × 20	7-1/2 × 19-1/2	146.3	475.3	4634	182.8	685.5	25.39	30.47	35.55	40.63	45.70	50.78
8 × 22	7-1/2 × 21-1/2	161.3	577.8	6211	201.6	755.9	27.99	33.59	39.19	44.79	50.39	55.99
8 × 24	7-1/2 × 23-1/2	176.3	690.3	8111	220.3	826.2	30.60	36.72	42.84	48.96	55.08	61.20
10 × 10	9-1/2 × 9-1/2	90.25	142.9	678.8	142.9	678.8	15.67	18.80	21.94	25.07	28.20	31.34
10 × 12	9-1/2 × 11-1/2	109.3	209.4	1204	173.0	821.7	18.97	22.76	26.55	30.35	34.14	37.93
10 × 14	9-1/2 × 13-1/2	128.3	288.6	1948	203.1	964.5	22.27	26.72	31.17	35.63	40.08	44.53
10 × 16	9-1/2 × 15-1/2	147.3	380.4	2948	233.1	1107	25.56	30.68	35.79	40.90	46.02	51.13
10 × 18	9-1/2 × 17-1/2	166.3	484.9	4243	263.2	1250	28.86	34.64	40.41	46.18	51.95	57.73
10 × 20	9-1/2 × 19-1/2	185.3	602.1	5870	293.3	1393	32.16	38.59	45.03	51.46	57.89	64.32
10 × 22	9-1/2 × 21-1/2	204.3	731.9	7868	323.4	1536	35.46	42.55	49.64	56.74	63.83	70.92
10 × 24	9-1/2 × 23-1/2	223.3	874.4	10270	353.5	1679	38.76	46.51	54.26	62.01	69.77	77.52

12 × 12	11-1/2 × 11-1/2	132.3	253.5	1458	253.5	1458	22.96	27.55	32.14	36.74	41.33	45.92
12 × 14	11-1/2 × 13-1/2	155.3	349.3	2358	297.6	1711	26.95	32.34	37.73	43.13	48.52	53.91
12 × 16	11-1/2 × 15-1/2	178.3	460.5	3569	341.6	1964	30.95	37.14	43.32	49.51	55.70	61.89
12 × 18	11-1/2 × 17-1/2	201.3	587.0	5136	385.7	2218	34.94	41.93	48.91	55.90	62.89	69.88
12 × 20	11-1/2 × 19-1/2	224.3	728.8	7106	429.8	2471	38.93	46.72	54.51	62.29	70.08	77.86
12 × 22	11-1/2 × 21-1/2	247.3	886.0	9524	473.9	2725	42.93	51.51	60.10	68.68	77.27	85.85
12 × 24	11-1/2 × 23-1/2	270.3	1058	12440	518.0	2978	46.92	56.30	65.69	75.07	84.45	93.84
14 × 14	13-1/2 × 13-1/2	182.3	410.1	2768	410.1	2768	31.64	37.97	44.30	50.63	56.95	63.28
14 × 16	13-1/2 × 15-1/2	209.3	540.6	4189	470.8	3178	36.33	43.59	50.86	58.13	65.39	72.66
14 × 18	13-1/2 × 17-1/2	236.3	689.1	6029	531.6	3588	41.02	49.22	57.42	65.63	73.83	82.03
14 × 20	13-1/2 × 19-1/2	263.3	855.6	8342	592.3	3998	45.70	54.84	63.98	73.13	82.27	91.41
14 × 22	13-1/2 × 21-1/2	290.3	1040	11180	653.1	4408	50.39	60.47	70.55	80.63	90.70	100.8
14 × 24	13-1/2 × 23-1/2	317.3	1243	14600	713.8	4818	55.08	66.09	77.11	88.13	99.14	110.2
16 × 16	15-1/2 × 15-1/2	240.3	620.6	4810	620.6	4810	41.71	50.05	58.39	66.74	75.08	83.42
16 × 18	15-1/2 × 17-1/2	271.3	791.1	6923	700.7	5431	47.09	56.51	65.93	75.35	84.77	94.18
16 × 20	15-1/2 × 19-1/2	302.3	982.3	9578	780.8	6051	52.47	62.97	73.46	83.96	94.45	104.9
16 × 22	15-1/2 × 21-1/2	333.3	1194	12840	860.9	6672	57.86	69.43	81.00	92.57	104.1	115.7
16 × 24	15-1/2 × 23-1/2	364.3	1427	16760	941.0	7293	63.24	75.89	88.53	101.2	113.8	126.5
18 × 18	17-1/2 × 17-1/2	306.3	893.2	7816	893.2	7816	53.17	63.80	74.44	85.07	95.70	106.3
18 × 20	17-1/2 × 19-1/2	341.3	1109	10810	995.3	8709	59.24	71.09	82.94	94.79	106.6	118.5
18 × 22	17-1/2 × 21-1/2	376.3	1348	14490	1097	9602	65.32	78.39	91.45	104.5	117.6	130.6
18 × 24	17-1/2 × 23-1/2	411.3	1611	18930	1199	10500	71.40	85.68	99.96	114.2	128.5	142.8
20 × 20	19-1/2 × 19-1/2	380.3	1236	12050	1236	12050	66.02	79.22	92.42	105.6	118.8	132.0
20 × 22	19-1/2 × 21-1/2	419.3	1502	16150	1363	13280	72.79	87.34	101.9	116.5	131.0	145.6
20 × 24	19-1/2 × 23-1/2	458.3	1795	21090	1489	14520	79.56	95.47	111.4	127.3	143.2	159.1
22 × 22	21-1/2 × 21-1/2	462.3	1656	17810	1656	17810	80.25	96.30	112.4	128.4	144.5	160.5
22 × 24	21-1/2 × 23-1/2	505.3	1979	23250	1810	19460	87.72	105.3	122.8	140.3	157.9	175.4
24 × 24	23-1/2 × 23-1/2	552.3	2163	25420	2163	25420	95.88	115.1	134.2	153.4	172.6	191.8

From National Design Specification for Wood Construction 1991

3. Posts and timbers: Cross section is approximately 5 × 5 in.
    square or larger, and *width* > *thickness* + 2 in. (not more)
    Decking: 2 in. ≤ thickness ≤ 4 in. with load applied to wide
    face of board

All sizes referred to in the previous classification are the *nom-inal*, or stated, sizes. However, most lumber is called *dressed* lum-ber, which means the members are surfaced to a standard net size. Structural computations to determine the required size of mem-bers are based on the net dimensions (actual sizes), not the nomi-nal size. Sizes of members is further discussed in Section 3.3.2.

3.3.2   Sizes of Structural Lumber

Most structural lumber is called dressed lumber. In other words, the lumber is surfaced to the standard net size, which is less than the nominal, or stated, size. This is shown in Figure 3.3.

Dressed lumber is used in many structural applications. How-ever, some architectural applications may call for larger members that have a different texture. Such members are commonly *rough-sawn* to dimensions that are close to the standard net size. The cross-sectional dimensions of these timbers is about ⅛ in. larger

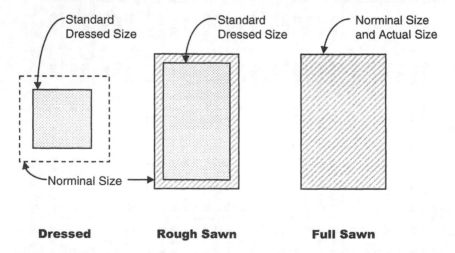

Dressed   **Rough Sawn**   **Full Sawn**

Figure 3.3   Example of sizes of structural lumber.

than the standard dressed size. A less common method of obtaining a rough surface is to specify *full-sawn lumber*. Since rough-sawn and full-sawn lumber are not frequently used, their cross-sectional properties are not included in the NDS.

Below is an example of the differences between nominal, dressed, rough-sawn, and full-sawn sizes of lumber. Consider an 8 × 12 member (nominal size = 8 × 12 in.):

1. *Dressed lumber*: Standard net size 7½ × 11½ in.
2. *Rough-sawn lumber*: Approximate size 7⅝ × 11⅝ in.
3. *Full-sawn lumber*: Minimum size 8 × 12 in. (generally not available).

3.3.3 Mechanical Properties of Lumber

The mechanical properties that will be used in the design of formwork are compression parallel to grain (F_c), compression perpendicular to grain $(F_{c\perp})$, tension parallel to grain (F_t), and tension perpendicular to grain $(F_{t\perp})$ Figure 3.4 helps clarify the direction of forces which produce these different types of stresses.

3.3.4 Design Values of Mechanical Properties

Design values for the different types of stresses are dependent on the type of lumber. The design values given in these tables are to be adjusted to fit the conditions under which the structure will be used. Tables 3.3 through 3.6 give the design values along with its adjustment factors that are specified by NDS for dimension lumber, southern pine dimension lumber, timber (5 × 5 in. and larger) and decking. Table 3.3a through 3.3d gives the design values along with its adjustment factors for all species except Southern Pine. Design values for Southern Pine are shown in Tables 3.4a through 3.4d and Table 3.5.

Size Factor

Stresses parallel to grain for visually graded dimension lumber should be multiplied by the size factors provided in Tables 3.3a and 3.4a.

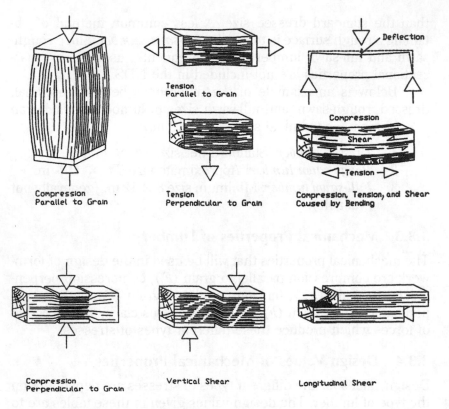

Figure 3.4 Forces and directions of grains.

When the depth d of the beam, stringer, post, or timber exceeds 12 in., the tabulated design value F_b shall be multiplied by the following size factor:

$$C_F = \left(\frac{12.0}{d}\right)^{1/9}$$

Effect of Moisture

Dry service conditions are those in which the moisture content during the use of the member will not be more than 19 percent,

Table 3.3 Design Values For Visually Graded Dimension Lumber

Species and commercial grade	Size classification	Design values in pounds per square inch (psi)						Grading Rules Agency
		Bending F_b	Tension parallel to grain F_t	Shear parallel to grain F_v	Compression perpendicular to grain $F_{c\perp}$	Compression parallel to grain F_c	Modulus of Elasticity E	
DOUGLAS FIR-LARCH								
Select Structural	2"–4" thick	1450	1000	95	625	1700	1,900,000	
No. 1 and Better		1150	775	95	625	1500	1,800,000	
No. 1		1000	675	95	625	1450	1,700,000	
No. 2	2" & wider	875	575	95	625	1300	1,600,000	WCLIB*
No. 3		500	325	95	625	750	1,400,000	WWPA**
Stud		675	450	95	625	825	1,400,000	
Construction	2"–4" thick	1000	650	95	625	1600	1,500,000	
Standard		550	375	95	625	1350	1,400,000	
Utility	2"–4" wide	275	175	95	625	875	1,300,000	
DOUGLAS FIR-LARCH (NORTH)								
Select Structural	2"–4" thick	1300	800	95	625	1900	1,900,000	
No. 1/No. 2		825	500	95	625	1350	1,600,000	
No. 3	2" & wider	475	300	95	625	775	1,400,000	NLGA***
Stud		650	375	95	625	850	1,400,000	
Construction	2"–4" thick	950	575	95	625	1750	1,500,000	
Standard		525	325	95	625	1400	1,400,000	
Utility	2"–4" wide	250	150	95	625	925	1,300,000	

* West Coast Lumber Inspection Bureau
** Western Wood Products Association
*** Northeastern Lumber Grading Agency
From National Design Specification for Wood Construction 1991

Table 3.3 Continued

Species and commercial grade	Size classification	Design values in pounds per square inch (psi)						Grading Rules Agency
		Bending F_b	Tension parallel to grain F_t	Shear parallel to grain F_v	Compression perpendicular to grain $F_{c\perp}$	Compression parallel to grain F_c	Modulus of Elasticity E	
DOUGLAS FIR-SOUTH								
Select Structural	2"–4" thick	1300	875	90	520	1550	1,400,000	
No. 1		900	600	90	520	1400	1,300,000	
No. 2		825	525	90	520	1300	1,200,000	WWPA
No. 3	2" & wider	475	300	90	520	750	1,100,000	
Stud		*650*	*425*	*90*	*520*	*825*	*1,100,000*	
Construction	*2"–4" thick*	*925*	*600*	*90*	*520*	*1550*	*1,200,000*	
Standard		*525*	*350*	*90*	*520*	*1300*	*1,100,000*	
Utility	*2"–4" wide*	*250*	*150*	*90*	*520*	*875*	*1,000,000*	
EASTERN HEMLOCK-TAMARACK								
Select Structural	2"–4" thick	1250	575	85	555	1200	1,200,000	
No. 1		775	350	85	555	1000	1,100,000	
No. 2		575	275	85	555	825	1,100,000	NELMA*
No. 3	2" & wider	350	150	85	555	475	900,000	NSLB**
Stud		*450*	*200*	*85*	*555*	*525*	*900,000*	
Construction	*2"–4" thick*	*675*	*300*	*85*	*555*	*1050*	*1,000,000*	
Standard		*375*	*175*	*85*	*555*	*850*	*900,000*	
Utility	*2"–4" wide*	*175*	*75*	*85*	*555*	*550*	*800,000*	

* Northeastern Lumber Manufacturers Association
** Northern Softwood Lumber Bureau

EASTERN SOFTWOODS

Grade	Size						Agency
Select Structural	2"-4" thick	1250	575	70	335	1,200,000	
No. 1		775	350	70	335	1,100,000	
No. 2		575	275	70	335	1,100,000	
No. 3	2" & wider	350	150	70	335	900,000	NELMA
Stud		*450*	*200*	*70*	*335*	*900,000*	*NSLB*
Construction	*2"-4" thick*	*675*	*300*	*70*	*335*	*1,000,000*	
Standard		*375*	*175*	*70*	*335*	*900,000*	
Utility	*2"-4" wide*	*175*	*75*	*70*	*335*	*800,000*	

EASTERN WHITE PINE

Grade	Size						Agency
Select Structural	2"-4" thick	1250	575	70	350	1,200,000	
No. 1		775	350	70	350	1,100,000	
No. 2		575	275	70	350	1,100,000	
No. 3	2" & wider	350	150	75	350	900,000	NELMA
Stud		*450*	*200*	*70*	*350*	*900,000*	*NSLB*
Construction	*2"-4" thick*	*675*	*300*	*70*	*350*	*1,000,000*	
Standard		*375*	*175*	*70*	*350*	*900,000*	
Utility	*2"-4" wide*	*175*	*75*	*70*	*350*	*800,000*	

HEM-FIR

Grade	Size						Agency
Select Structural	2"-4" thick	1400	900	75	405	1,600,000	
No. 1 & Btr		1050	700	75	405	1,500,000	
No. 1		950	600	75	405	1,500,000	
No. 2	2" & wider	850	500	75	405	1,300,000	WCLIB
No. 3		500	300	75	405	1,200,000	WWPA
Stud	*2"-4" thick*	*675*	*400*	*75*	*405*	*1,200,000*	
Construction		*975*	*575*	*75*	*405*	*1,300,000*	
Standard		*550*	*325*	*75*	*405*	*1,200,000*	
Utility	*2"-4" wide*	*250*	*150*	*75*	*405*	*1,100,000*	

Table 3.3 Continued

Species and commercial grade	Size classification	Design values in pounds per square inch (psi)						Grading Rules Agency
		Bending F_b	Tension parallel to grain F_t	Shear parallel to grain F_v	Compression perpendicular to grain $F_{c\perp}$	Compression parallel to grain F_c	Modulus of Elasticity E	
HEM-FIR (NORTH)								
Select Structural	2″–4″ thick	1300	775	75	370	1650	1,700,000	NLGA
No. 1/No. 2		1000	550	75	370	1450	1,600,000	
No. 3	2″ & wider	575	325	75	370	850	1,400,000	
Stud		*775*	*425*	*75*	*370*	*925*	*1,400,000*	
Construction	*2″–4″ thick*	*1150*	*625*	*75*	*370*	*1750*	*1,500,000*	
Standard		*625*	*350*	*75*	*370*	*1500*	*1,400,000*	
Utility	*2″–4″ wide*	*300*	*175*	*75*	*370*	*975*	*1,300,000*	
MIXED MAPLE								
Select Structural	2″–4″ thick	1000	600	100	620	875	1,300,000	NELMA
No. 1		725	425	100	620	700	1,200,000	
No. 2		700	425	100	620	550	1,100,000	
No. 3	2″ & wider	400	250	100	620	325	1,000,000	
Stud		*550*	*325*	*100*	*620*	*350*	*1,000,000*	
Construction	*2″–4″ thick*	*800*	*475*	*100*	*620*	*725*	*1,100,000*	
Standard		*450*	*275*	*100*	*620*	*575*	*1,000,000*	
Utility	*2″–4″ wide*	*225*	*125*	*100*	*620*	*375*	*900,000*	

MIXED OAK

Grade	Size						
Select Structural	2"-4" thick	1150	675	85	800	1000	1,100,000
No. 1		825	500	85	800	825	1,000,000
No. 2		800	475	85	800	625	900,000
No. 3	2" & wider	475	275	85	800	375	800,000 NELMA
Stud		625	375	85	800	400	800,000
Construction	2"-4" thick	925	550	85	800	850	900,000
Standard		525	300	85	800	650	800,000
Utility	2"-4" wide	250	150	85	800	425	800,000

NORTHERN RED OAK

Grade	Size						
Select Structural	2"-4" thick	1400	800	110	885	1150	1,400,000
No. 1		1000	575	110	885	925	1,400,000
No. 2		975	575	110	885	725	1,300,000
No. 3	2" & wider	550	325	110	885	425	1,200,000 NELMA
Stud		750	450	110	885	450	1,200,000
Construction	2"-4" thick	1100	650	110	885	975	1,200,000
Standard		625	350	110	885	750	1,100,000
Utility	2"-4" wide	300	175	110	885	500	1,000,000

NORTHERN SPECIES

Grade	Size						
Select Structural	2"-4" thick	950	450	65	350	1100	1,100,000
No. 1/No. 2		575	275	65	350	825	1,100,000
No. 3	2" & wider	350	150	65	350	475	1,000,000 NLGA
Stud		450	200	65	350	525	1,000,000
Construction	2"-4" thick	675	300	65	350	1050	1,000,000
Standard		375	175	65	350	850	900,000
Utility	2"-4" wide	175	75	65	350	550	900,000

Table 3.3 Continued

Species and commercial grade	Size classification	Design values in pounds per square inch (psi)						Grading Rules Agency
		Bending F_b	Tension parallel to grain F_t	Shear parallel to grain F_v	Compression perpendicular to grain $F_{c\perp}$	Compression parallel to grain F_c	Modulus of Elasticity E	
NORTHERN WHITE CEDAR								
Select Structural	2"–4" thick	775	450	60	370	750	800,000	
No. 1		575	325	60	370	600	700,000	
No. 2		550	325	60	370	475	700,000	
No. 3	2" & wider	325	175	60	370	275	600,000	NELMA
Stud		*425*	*250*	*60*	*370*	*300*	*600,000*	
Construction	*2"–4" thick*	*625*	*375*	*60*	*370*	*625*	*700,000*	
Standard		*350*	*200*	*60*	*370*	*475*	*600,000*	
Utility	*2"–4" wide*	*175*	*100*	*60*	*370*	*325*	*600,000*	
RED MAPLE								
Select Structural	2"–4" thick	1300	750	105	615	1100	1,700,000	
No. 1		925	550	105	615	900	1,600,000	
No. 2		900	525	105	615	700	1,500,000	
No. 3	2" & wider	525	300	105	615	400	1,300,000	NELMA
Stud		*700*	*425*	*105*	*615*	*450*	*1,300,000*	
Construction	*2"–4" thick*	*1050*	*600*	*105*	*615*	*925*	*1,400,000*	
Standard		*575*	*325*	*105*	*615*	*725*	*1,300,000*	
Utility	*2"–4" wide*	*275*	*150*	*105*	*615*	*475*	*1,200,000*	

RED OAK

Grade	Size							
Select Structural	2"–4" thick	1150	675	85	820	1000	1,400,000	
No. 1		825	500	85	820	825	1,300,000	
No. 2		800	475	85	820	625	1,200,000	
No. 3	2" & wider	475	275	85	820	375	1,100,000	NELMA
Stud		625	375	85	820	400	1,100,000	
Construction	2"–4" thick	925	550	85	820	850	1,200,000	
Standard		525	300	85	820	650	1,100,000	
Utility	2"–4" wide	250	150	85	820	425	1,000,000	

REDWOOD

Grade	Size							
Clear Structural		1750	1000	145	650	1850	1,400,000	
Select Structural		1350	800	80	650	1500	1,400,000	
Select Structural, open grain		1100	625	85	425	1100	1,100,000	
No. 1	2"–4" thick	975	575	80	650	1200	1,300,000	
No. 1, open grain		775	450	80	425	900	1,100,000	
No. 2	2" & wider	925	525	80	650	950	1,200,000	RIS*
No. 2, open grain		725	425	80	425	700	1,000,000	
No. 3		525	300	80	650	550	1,100,000	
No. 3, open grain		425	250	80	425	400	900,000	
Stud		575	325	80	425	450	900,000	
Construction	2"–4" thick	825	475	80	425	925	900,000	
Standard		450	275	80	425	725	900,000	
Utility	2"–4" wide	225	125	80	425	475	800,000	

* Redwood Inspection Service

Table 3.3 Continued

Species and commercial grade	Size classification	Design values in pounds per square inch (psi)						Grading Rules Agency
		Bending F_b	Tension parallel to grain F_t	Shear parallel to grain F_v	Compression perpendicular to grain $F_{c\perp}$	Compression parallel to grain F_c	Modulus of Elasticity E	
SPRUCE-PINE-FIR								
Select Structural	2″–4″ thick	1250	675	70	425	1400	1,500,000	
No. 1/No. 2		875	425	70	425	1100	1,400,000	NLGA
No. 3	2″ & wider	500	250	70	425	625	1,200,000	
Stud		675	325	70	425	675	1,200,000	
Construction	2″–4″ thick	975	475	70	425	1350	1,300,000	
Standard		550	275	70	425	1100	1,200,000	
Utility	2″–4″ wide	250	125	70	425	725	1,100,000	
SPRUCE-PINE-FIR (SOUTH)								
Select Structural	2″–4″ thick	1300	575	70	335	1200	1,300,000	
No. 1		850	400	70	335	1050	1,200,000	NELMA
No. 2		750	325	70	335	975	1,100,000	NSLB
No. 3	2″ & wider	425	200	70	335	550	1,000,000	WCLIB
Stud		575	250	70	335	600	1,000,000	WWPA
Construction	2″–4″ thick	850	375	70	335	1200	1,000,000	
Standard		475	225	70	335	1000	900,000	
Utility	2″–4″ wide	225	100	70	335	650	900,000	

WESTERN CEDARS

Select Structural	2"–4" thick	1000	600	75	425	1000	1,100,000 WCLIB
No. 1		725	425	75	425	825	1,000,000 WWPA
No. 2		700	425	75	425	650	1,000,000
No. 3	2" & wider	400	250	75	425	375	900,000
Stud		550	325	75	425	400	900,000
Construction	2"–4" thick	800	475	75	425	850	900,000
Standard		450	275	75	425	650	800,000
Utility	2"–4" wide	225	125	75	425	425	800,000

WESTERN WOODS

Select Structural	2"–4" thick	875	400	70	335	1050	1,200,000 WCLIB
No. 1		650	300	70	335	925	1,100,000 WWPA
No. 2		650	275	70	335	875	1,000,000
No. 3	2" & wider	375	175	70	335	500	900,000
Stud		500	225	70	335	550	900,000
Construction	2"–4" thick	725	325	70	335	1050	1,000,000
Standard		400	175	70	335	900	900,000
Utility	2"–4" wide	200	75	70	335	600	800,000

Table 3.3a Size Adjustment Factor C_F for Visually Graded Dimension Lumber

Grades	Width (in.)	F_b		F_t	F_c
		Thickness 2 and 3 in.	Thickness 4 in.		
Select Structural, No. 1 and Better No. 2 No. 3	2, 3, and 4	1.5	1.5	1.5	1.15
	5	1.4	1.4	1.4	1.1
	6	1.3	1.3	1.3	1.1
	8	1.2	1.3	1.2	1.05
	10	1.1	1.2	1.1	1.0
	12	1.0	1.1	1.0	1.0
	14 and wider	0.9	1.0	0.9	0.9
Stud	2, 3, and 4	1.1	1.1	1.1	1.05
	5 and 6	1.0	1.0	1.0	1.0
Construction and Standard	2, 3, and 4	1.0	1.0	1.0	1.0
Utility	4	1.0	1.0	1.0	1.0
	2 and 3	0.4	—	0.4	0.6

Table 3.3b Flat Use Factor C_{fu}

Width in.	Thickness (in.)	
	2 and 3	4
2 and 3	1.0	—
4	1.1	1.0
5	1.1	1.05
6	1.15	1.05
8	1.15	1.05
10 and wider	1.2	1.1

Table 3.3c Wet Service Adjustment Factor C_M

Type of stress	F_b	F_t	F_v	$F_{c\perp}$	F_c	E
C_M	0.85*	1.0	0.97	0.67	0.8†	0.9

* When $(F_b)(C_F) \leq 1150$ psi, $C_M = 1.0$.
† When $(F_c)(C_F) \leq 750$ psi, $C_M = 1.0$.

Table 3.3d Shear Stress Factor C_H

Length of split on wide face of 2-in. (nominal) lumber	C_H	Length of split on wide face of 3-in. (nominal) and thicker lumber	C_H	Size of shake* in 2-in. (nominal) and thicker lumber	C_H
No split	2.00	No split	2.00	No shake	2.00
½ × wide face	1.67	½ × narrow face	1.67	⅙ × narrow face	1.67
¾ × wide face	1.50	¾ × narrow face	1.50	¼ × narrow face	1.50
1 × wide face	1.33	1 × narrow face	1.33	⅓ × narrow face	1.33
1-½ × wide face or more	1.00	1-½ × narrow face or more	1.00	½ × narrow face or more	1.00

* Shake is measured at the end between lines enclosing the shake and perpendicular to the loaded surface.

Table 3.4 Base Design Values For Visually Graded Mixed Southern Pine Dimension Lumber

Species and commercial grade	Size classification	Bending F_b	Tension parallel to grain F_t	Shear parallel to grain F_v	Compression perpendicular to grain $F_{c\perp}$	Compression parallel to grain F_c	Modulus of Elasticity E	Grading Rules Agency
MIXED SOUTHERN PINE								
Select Structural	2″–4″ thick	2050	1200	100	565	1800	1,600,000	
No. 1		1450	875	100	565	1650	1,500,000	
No. 2		1300	775	90	565	1650	1,400,000	
No. 3	2″–4″ wide	750	450	90	565	950	1,200,000	
Stud		775	450	90	565	950	1,200,000	SPIB*
Construction	2″–4″ thick	1000	600	100	565	1700	1,300,000	
Standard		550	325	90	565	1450	1,200,000	
Utility	4″ wide	275	150	90	565	950	1,100,000	

Select Structural	2"-4" thick	1850	1100	90	565	1700	1,600,000
No. 1		1300	750	90	565	1550	1,500,000
No. 2		1150	675	90	565	1550	1,400,000
No. 3	5"-6" wide	675	400	90	565	875	1,200,000
Stud		675	400	90	565	875	1,200,000
Select Structural	2"-4" thick	1750	1000	90	565	1600	1,600,000
No. 1		1200	700	90	565	1450	1,500,000
No. 2	8" wide	1050	625	90	565	1450	1,400,000
No. 3		625	375	90	565	850	1,200,000
Select Structural	2"-4" thick	1500	875	90	565	1600	1,600,000
No. 1		1050	600	90	565	1450	1,500,000
No. 2	10" wide	925	550	90	565	1450	1,400,000
No. 3		565	325	90	565	825	1,200,000
Select Structural	2"-4" thick	1400	825	90	565	1550	1,600,000
No. 1		975	575	90	565	1400	1,500,000
No. 2	12" wide	875	525	90	565	1400	1,400,000
No. 3		500	300	90	565	800	1,200,000

* Southern Pine Inspection Bureau

Table 3.4a Size Adjustment Factor C_F for Southern Pine Including Mixed

	For dimension lumber 4 in. thick, 8 in. and wider	For 12 in. and wider lumber
C_F	1.1	0.9

Table 3.4b Wet Service Adjustment Factor C_M

Type of stress	F_b	F_t	F_v	$F_{c\perp}$	F_c	E
C_M	0.85*	1.0	0.97	0.67	0.8†	0.9

* When $(F_b)(C_F) \leq 1150$ psi, $C_M = 1.0$.
† When $(F_c) \leq 750$ psi, $C_M = 1.0$.

Table 3.4c Flat Use Factor C_{fu}

Width (in.)	Thickness (in.)	
	2 and 3	4
2 and 3	1.0	—
4	1.1	1.0
5	1.1	1.05
6	1.15	1.05
8	1.15	1.05
10 and wider	1.2	1.1

Table 3.4d Shear Stress Factor C_H

Length of split on wide face of 2-in. (nominal) lumber	C_H	Length of split on wide face of 3-in. (nominal) and thicker lumber	C_H	Size of shake* in 2-in. (nominal) and thicker lumber	C_H
No split	2.00	No split	2.00	No shake	2.00
½ × wide face	1.67	½ × narrow face	1.67	⅙ × narrow face	1.67
¾ × wide face	1.50	¾ × narrow face	1.50	¼ narrow face	1.50
1 × wide face	1.33	1 × narrow face	1.33	⅓ narrow face	1.33
1-½ × wide face or more	1.0	1-½ × narrow face or more	1.00	½ × narrow face or more	1.00

* Shake is measured at the end between lines enclosing the shake and perpendicular to the loaded surface.

Table 3.5 Base Design Values for Visually Graded Southern Pine Dimension Lumber

Species and commercial grade	Size classification	Design values in pounds per square inch (psi)						Grading Rules Agency
		Bending F_b	Tension parallel to grain F_t	Shear parallel to grain F_v	Compression perpendicular to grain $F_{c\perp}$	Compression parallel to grain F_c	Modulus of Elasticity E	
SOUTHERN PINE								
Dense Select Structural		3050	1650	100	660	2250	1,900,000	
Select Structural		2850	1600	100	565	2100	1,800,000	
Non-Dense Select Structural		2650	1350	100	480	1950	1,700,000	
No. 1 Dense	2"-4" thick	2000	1100	100	660	2000	1,800,000	
No. 1		1850	1050	100	565	1850	1,700,000	SPIB*
No. 1 Non-Dense	2"-4" wide	1700	900	100	480	1700	1,600,000	
No. 2 Dense		1700	875	90	660	1850	1,700,000	
No. 2		1500	825	90	565	1650	1,600,000	
No. 2 Non-Dense		1350	775	90	480	1600	1,400,000	
No. 3		850	475	90	565	975	1,400,000	
Stud		875	500	90	565	975	1,400,000	
Construction	2"-4" thick	1100	625	100	565	1800	1,500,000	
Standard		625	350	90	565	1500	1,300,000	
Utility	4" wide	300	175	90	565	975	1,300,000	
Dense Select Structural		2700	1500	90	660	2150	1,900,000	
Select Structural		2550	1400	90	565	2000	1,800,000	
Non-Dense Select Structural		2350	1200	90	480	1850	1,700,000	
No. 1 Dense		1750	950	90	660	1900	1,800,000	

Grade	Size						SPIB
No. 1	2″–4″ thick, 5″–6″ wide	1650	900	90	565	1750	1,700,000
No. 1 Non-Dense		1500	800	90	480	1600	1,600,000
No. 2 Dense		1450	775	90	660	1750	1,700,000
No. 2		1250	725	90	565	1600	1,600,000
No. 2 Non-Dense		1150	675	90	480	1500	1,400,000
No. 3		750	425	90	565	925	1,400,000
Stud		775	425	90	565	925	1,400,000
Dense Select Structural	2″–4″ thick, 8″ wide	2450	1350	90	660	2050	1,900,000
Select Structural		2300	1300	90	565	1900	1,800,000
Non-Dense Select Structural		2100	1100	90	480	1750	1,700,000
No. 1 Dense		1650	875	90	660	1800	1,800,000
No. 1		1500	825	90	565	1650	1,700,000
No. 1 Non-Dense		1350	725	90	480	1550	1,600,000
No. 2 Dense		1400	675	90	660	1700	1,700,000
No. 2		1200	650	90	565	1550	1,600,000
No. 2 Non-Dense		1100	600	90	480	1450	1,400,000
No. 3		700	400	90	565	875	1,400,000
Dense Select Structural	2″–4″ thick, 10″ wide	2150	1200	90	660	2000	1,900,000
Select Structural		2050	1100	90	565	1850	1,800,000
Non-Dense Select Structural		1850	950	90	480	1750	1,700,000
No. 1 Dense		1450	775	90	660	1750	1,800,000
No. 1		1300	725	90	565	1600	1,700,000
No. 1 Non-Dense		1200	650	90	480	1500	1,600,000
No. 2 Dense		1200	625	90	660	1650	1,700,000
No. 2		1050	575	90	565	1500	1,600,000
No. 2 Non-Dense		950	550	90	480	1400	1,400,000
No. 3		600	325	90	565	850	1,400,000

Table 3.5 Continued

Species and commercial grade	Size classification	Design values in pounds per square inch (psi)						Grading Rules Agency
		Bending F_b	Tension parallel to grain F_t	Shear parallel to grain F_v	Compression perpendicular to grain $F_{c\perp}$	Compression parallel to grain F_c	Modulus of Elasticity E	
Dense Select Structural		2050	1100	90	660	1950	1,900,000	
Select Structural		1900	1050	90	565	1800	1,800,000	
Non-Dense Select Structural		1750	900	90	480	1700	1,700,000	
No. 1 Dense	2"–4" thick	1350	725	90	660	1700	1,800,000	
No. 1		1250	675	90	565	1600	1,700,000	
No. 1 Non-Dense	12" wide[4]	1150	600	90	480	1500	1,600,000	
No. 2 Dense		1150	575	90	660	1600	1,700,000	
No. 2		975	550	90	565	1450	1,600,000	
No. 2 Non-Dense		900	525	90	480	1350	1,400,000	
No. 3		575	325	90	565	825	1,400,000	
SOUTHERN PINE (Dry service conditions—19% or less moisture content)								
Dense Structural 86	2-1/2"–4" thick	2600	1750	155	660	2000	1,800,000	SPIB
Dense Structural 72		2200	1450	130	660	1650	1,800,000	
Dense Structural 65	2" & wider	2000	1300	115	660	1500	1,800,000	
SOUTHERN PINE (Wet service conditions)								
Dense Structural 86	2-1/2"–4" thick	2100	1400	145	440	1300	1,600,000	SPIB
Dense Structural 72		1750	1200	120	440	1100	1,600,000	
Dense Structural 65	2-1/2" & wider	1600	1050	110	440	1000	1,600,000	

* Southern Pine Inspection Bureau

From National Design Specification for Wood Construction 1991

Table 3.5a Wet Service Adjustment Factor C_M

Type of stress	F_b	F_t	F_v	$F_{c\perp}$	F_c	E
C_M	1.00	1.0	1.00	0.67	0.91	1.00

Table 3.5b Shear Stress Factor C_H

Length of split on wide face of 5-in. (nominal) and thicker lumber	C_H	Size of shake* in 5-in. (nominal) and thicker lumber	C_H
No split	2.00	No shake	2.00
$\tfrac{1}{2} \times$ narrow width	1.67	$\tfrac{1}{6} \times$ narrow face	1.67
$\tfrac{3}{4} \times$ narrow width	1.50	$\tfrac{1}{4} \times$ narrow face	1.50
$1 \times$ narrow width	1.33	$\tfrac{1}{3} \times$ narrow face	1.33
$1\tfrac{1}{2} \times$ narrow width	1.00	$\tfrac{1}{2} \times$ narrow face or more	1.00

* Shake is measured at the end between lines enclosing the shake and perpendicular to the loaded surface.

Table 3.6 Design Values for Visually Graded Decking

Species and commercial grade	Size classification	Design values in pounds per square inches		
		F_b	$F_{c\perp}$	E
Douglas Fir-Larch				
Select Dex Commercial Dex	2–4 in. thick 6–8 in. wide	1750.0 1450.0	625.0 625.0	1,800,000 1,700,000
Hem-Fir				
Select Dex Commercial Dex	2–4 in. thick 6–8 in. wide	1400.0 1150.0	405.0 405.0	1,500,000 1,400,000
Redwood				
Select, Close Grain Select Commercial	2 in. thick 6 in. and wider	1850.0 1450.0 1200.0	— — —	1,400,000 1,100,000 1,000,000
Deck heart and Deck common	2 in. thick, 4 in. wide	400.0	420.0	900,000
	2 in. thick, 6 in. wide	700.0	420.0	900,000
Southern Pine *(Dry service conditions—19% or less moisture content)*				
Dense Standard Dense Select Select Dense Commercial Commercial	2–4 in. thick 2 in. and wider	2000.0 1650.0 1400.0 1650.0 1400.0	660.0 660.0 565.0 660.0 565.0	1,800,000 1,600,000 1,600,000 1,600,000 1,600,000
Southern Pine *(Wet service conditions)*				
Dense Standard Dense Select Select Dense Commercial Commercial	2½–4 in. thick 2 in. and wider	1600.0 1350.0 1150.0 1350.0 1150.0	440.0 440.0 375.0 440.0 375.0	1,600,000 1,400,000 1,400,000 1,400,000 1,400,000

From National Design Specification for Wood Construction 1991

Table 3.6a Wet Service Factor $(C_M)^*$

Type of stress	F_b	$F_{c\perp}$	E
C_M	0.85†	0.67	0.9

* For Southern Pine use tabulated design values for wet service conditions without further adjustment.
† When $(F_b)(C_F) \leq 1150$ psi, $C_M = 1.0$.

regardless of the moisture content of the member at the time of its manufacture. The design values (allowable design stresses) for lumber are applicable to members that are used under dry service conditions, such as in most covered structures. For lumber used under conditions where the moisture content of the wood will exceed 19 percent for an extended period of time, the design values of the member shall be multiplied by the wet service condition factor C_M, as is specified by the National Design Specification for Wood Construction (NDS). The wet service factor (<1.0) is used to decrease the allowable stresses to account for the weakening of the member due to the increase in its moisture content. An example of the wet service factor is given in Table 3.3c.

Also, moisture content adds additional weight to the lumber. NDS specifies that the following formula shall be used to determine the density (in lb./ft³) of wood.

$$\text{Density} = 62.4 \left[\frac{G}{1 + G(0.009)(\text{m.c.})} \right] \left[1 + \frac{\text{m.c.}}{100} \right]$$

where

G = specific gravity of wood based on weight and volume when oven-dry
m.c. = moisture content of wood, %

Load Duration Factor

Wood has the property of carrying a substantially greater maximum load for short duration than it can for long duration of load-

Table 3.7 Load Duration Factor (C_D)

Load duration	C_D	Typical design load
Permanent	0.9	Dead load
10 years	1.0	Occupancy live load
2 months	1.15	Snow load
7 days	**1.25**	**Construction load**
10 minutes	1.6	Wind/earthquake load
Impact	2.0	Impact load

From National Design Specification for Wood Construction 1991

ing. The tabulated design values given by NDS apply to normal load duration. Normal load duration is defined as the application of the full design load that fully stresses a member to its allowable design value for a cumulative period of approximately 10 years. Values for load duration factors are given in Table 3.7.

Temperature Factor

Wood increases in strength when cooled below normal temperatures and decreases in strength when heated. Prolonged heating to a temperature above 150°F may result in permanent loss of strength. Tabulated design values shall be multiplied by temperature factors C_t for a member that will experience sustained exposure to elevated temperature up to 150°F. These values are shown in Table 3.8.

Bearing Area Factor

Tabulated compression design values perpendicular to grain $F_{c\perp}$ apply to bearings of any length at the ends of the member, and to all bearings 6 in. or more in length at any other location. For bearings less than 6 in. and not nearer than 3 in. to the end of a member, the tabulated design values perpendicular to grain $F_{c\perp}$ shall be permitted to be multiplied by the bearing area factor C_b. Values of C_b are given in Table 3.9.

Table 3.8 Temperature Factor (C_t)

Design values	In service moisture content	C_t		
		$T \leq 100°F$	$100°F < T \leq 125°F$	$125°F < T \leq 150°F$
F_t, E	Wet or dry	1.0	0.9	0.9
F_b, F_v, F_c, and $F_{c\perp}$	Dry	1.0	0.8	0.7
	Wet	1.0	0.7	0.5

From National Design Specification for Wood Construction 1991

Table 3.9 Bearing Area Factor (C_b)*

l_b (in.)	0.5	1.0	1.5	2.0	3.0	4.0	6 or more
C_b	1.75	1.38	1.25	1.19	1.13	1.10	1.00

* For round bearing area such as washers, the bearing length l_b will be equal to the diameter.

From National Design Specification for Wood Construction 1991

3.4 PROPERTIES OF PLYWOOD

3.4.1 Exposure Durability Classification

Plywood is classified as interior or exterior. The classification is made on the basis of the resistance of the glue bond to moisture, which is affected by the adhesive used, the veneer grade, and the panel construction. Plywood is made in four exposure durability classifications: Exterior, Exposure 1, IMAGE (Exposure 2), and Interior. Exterior type is made with 100 percent waterproof glue.

3.4.2 Veneer Classification

Plywood is graded based on the appearance and defects in the veneers. Veneer is divided into the following five grades: *N grade* is free from defects, knots, and restricted patches and is suitable for work where natural finish is desirable such as cabinet work. *A grade* is smooth, free of knots, and paintable. N and A are considered the highest grade levels. *B grade* is similar to A grade except that knots, patches, and sanding defects may be found. *C grade* allows larger knots and knotholes; it is the lowest grade allowed in exterior-type plywood. *C-plugged* is a repaired or improved C grade. *D grade* may have larger knots, knotholes, and a number of repairs, holes, and sanding defects; this grade is not permitted in exterior panels.

3.4.3 Plywood Grades

Any combination of the above-mentioned grades are available as face and back surface for the plywood panel.

The grade system of classification includes:

- C-D Interior and C-C exterior (both are unsanded) sheathing panels are designated APA Rated Sheathing.
- Touch sanded underlayment panels $^{19}/_{32}$ in. or thicker are designated APA Rated Sturd-I-Floor.
- Rated sheathing may be modified by the terms structural I or II.
- Plywood panels conforming to this system are designated by a thickness, or span, rating without reference to veneer species. Span rating is written in the form $^{48}/_{24}$, where the numerator indicates the maximum recommended span in inches when used for roofing and the denominator indicates the maximum recommended span in inches when used for subroofing. These values are based on a 35 lb/ft^2 (170.94 kg/m^2) roof load and a 100 lb/ft^2 (488.24 kg/m^2) floor load. The floor load limitation is a deflection of L/360 where L represents the span.

3.4.4 Wood Species

The woods which may be used to manufacture plywood under U.S. Product Standard PS-183 are classified into five groups based on elastic modulus in bending and important strength properties. The group classification of a plywood panel is usually determined by the face and back veneer with the inner veneer allowed to be of a different group. For certain grades such as Marine and the Structural I grades, however, all plies are required to be of Group 1 species.

3.4.5 Direction of Face Grain

For greatest strength and stiffness, plywood should be installed with the face of grain perpendicular to the supports. Section properties parallel to the face grain of the plywood are based on panel construction which gives minimum values in this direction. Properties perpendicular to the face grain are based on a usually different panel construction, which gives minimum values in that direction (Figure 1.6).

For design purposes, calculations must take into account in which direction the stresses will be imposed in the panels. If stresses can be expected in both directions, then both the parallel and the perpendicular direction should be checked.

3.4.6   Grade Stress Level

The allowable stresses shown in Table 3.10 are divided into three different levels, which are related to the grade of the plywood. Research indicates that strength is directly related to the veneer grade and glue type. The derivation of the stress level is as follows:

- Bending, tension, and compression stresses depend on the grade of the veneer.
- Shear stresses do not depend on the veneer grade, but do vary with type of glue.
- Stiffness and bearing strength depend on species group, not veneer grade.

It should be noted that Table 3.10 should be read in conjunction with Tables 3.11 to 3.14.

3.4.7   Dry Conditions and Wet Conditions

The allowable stresses in the column titled "dry" in Table 3.14 apply to plywood under service conditions that are continuously dry. Dry conditions are defined as a moisture content of less than 16 percent. When the moisture content is 16 percent or greater, the allowable stresses under the column titled "wet" will be used.

3.4.8   Load Duration

Since plywood used in formwork has a maximum load duration of 7 days, the APA specifications allow an increase in the design values given in Table 3.14 by 25 percent.

Table 3.10 Key to Span Rating and Species Group

Thickness (in.)	Span Rating (APA RATED SHEATHING grade)							Span Rating (STURD-I-FLOOR grade)			
	12/0	16/0	20/0	24/0	32/16	40/20	48/24	16 oc	20 oc	24 oc	48 oc
5/16	4	3	1								
3/8			4	1							
15/32 & 1/2				4	1[1]						
19/32 & 5/8					4	1					
23/32 & 3/4						4	1				
7/8										3[2]	
1-1/8											1

KEY TO SPAN RATING AND SPECIES GROUP

For panels with "Span Rating" as across top, and thickness as at left, use stress for species group given in table.

(1) Thicknesses not applicable to APA RATED STURD-I-FLOOR
(2) For APA RATED STURD-I-FLOOR 24 oc, use Group 4 stresses.

From Plywood Design Specification by The American Plywood Association 1997.

Table 3.11 Guide to Use of Allowable Stress and Section Properties Tables
EXTERIOR APPLICATIONS

Plywood grade	Description and use	Typical trademarks	Veneer grade Face	Back	Inner	Common thicknesses	Grade stress level (Table 3)	Species group	Section property table
APA RATED SHEATHING EXT[3]	Unsanded sheathing grade with waterproof glue bond for wall, roof, subfloor and industrial applications such as pallet bins.	APA THE ENGINEERED WOOD ASSOCIATION RATED SHEATHING 48/24 23/32 INCH SIZED FOR SPACING EXTERIOR 000 PS 1-95 C-C PRP-108	C	C	C	5/16, 3/8, 15/32, 1/2, 19/32, 5/8, 23/32, 3/4	S-1[6]	See "Key to Span Rating"	Table 1 (unsanded)
APA STRUCTURAL I RATED SHEATHING EXT[3]	"Structural" is a modifier for this unsanded sheathing grade. For engineered applications in construction and industry where full Exterior-type panels are required. Structural I is made from Group 1 woods only.	APA THE ENGINEERED WOOD ASSOCIATION RATED SHEATHING STRUCTURAL I 24/0 39 INCH SIZED FOR SPACING EXPOSURE 1 000 PS 1-95 C-D PRP-108	C	C	C	5/16, 3/8, 15/32, 1/2, 19/32, 5/8, 23/32, 3/4	S-1[6]	Group 1	Table 2 (unsanded)
APA RATED STURD-I-FLOOR EXT[3]	For combination subfloor-underlayment where severe moisture conditions may be present, as in balcony decks. Possesses high concentrated and impact load resistance during construction and occupancy. Touch-sanded. (4) Available with tongue-and-groove edges. (5)	APA THE ENGINEERED WOOD ASSOCIATION RATED STURD-I-FLOOR 20 OC 19/32 INCH SIZED FOR SPACING EXTERIOR 000 PS 1-95 OC PLBBGEDPRP 108	C plugged	C	C	19/32, 5/8, 23/32, 3/4	S-2	See "Key to Span Rating"	Table 1 (touch-sanded)

Grade	Description	Typical Trademark	Face	Back	Inner Plies	Thickness	Surface	Species Group	Table
APA UNDERLAYMENT EXT and APA C-C-PLUGGED EXT	Underlayment for floor where severe moisture conditions may exist. Also for controlled atmosphere rooms and many industrial applications. Touch-sanded. Available with tongue-and-groove edges. (5)	APA THE ENGINEERED WOOD ASSOCIATION / C-C PLUGGED GROUP 2 EXTERIOR 000 PS 1-95	C plugged	C	C	1/2, 19/32, 5/8, 23/32, 3/4	S-2	As Specified	Table 1 (touch-sanded)
APA B-B PLYFORM CLASS I or II [2]	Concrete-form grade with high reuse factor. Sanded both sides, mill-oiled unless otherwise specified. Available in HDO. For refined design information on this special-use panel see APA Design/Construction Guide: Concrete Forming, Form No. V345. Design using values from this specification will result in a conservative design. (5)	APA THE ENGINEERED WOOD ASSOCIATION / PLYFORM B-B CLASS 1 EXTERIOR 000 PS 1-95	B	B	C	19/32, 5/8, 23/32, 3/4	S-2	Class I use Group 1; Class II use Group 3	Table 1 (sanded)
APA-MARINE EXT	Superior Exterior-type plywood made only with Douglas-fir or Western Larch. Special solid-core construction. Available with MDO or HDO face. Ideal for boat hull construction.	MARINE · A·A · EXT APA · 000 · PS 1-95	A or B	A or B	B	1/4, 3/8, 1/2, 5/8, 3/4	A face & back use S-1 B face or back use S-2	Group 1	Table 2 (sanded)

Table 3.11 Continued

Plywood grade	Description and use	Typical trademarks	Veneer grade			Common thicknesses	Grade stress level (Table 3)	Species group	Section property table
			Face	Back	Inner				
APA APPEARANCE GRADES EXT	Generally applied where a high quality surface is required. Includes APA, A-A, A-B, A-C, B-B, B-C, HDO and MDO EXT. (5)	APA THE ENGINEERED WOOD ASSOCIATION A-C GROUP 1 EXTERIOR 000 PS 1-95	B or better	C or better	C	1/4, 11/32, 3/8, 15/32, 1/2, 19/32, 5/8, 23/32, 3/4	A or C face & back use S-1[6] B face or back use S2	As Specified	Table 1 (sanded)

(1) When exterior glue is specified, i.e. Exposure 1, stress level 2 (S-2) should be used.
(2) Check local suppliers for availability before specifying Plyform Class II grade, as it is rarely manufactured.
(3) Properties and stresses apply only to APA RATED STURD-I-FLOOR and APA RATED SHEATHING manufactured entirely with veneers.
(4) APA RATED STURD-I-FLOOR 2-4-1 may be produced unsanded.
(5) May be available as Structural I. For such designation use Group 1 stresses and Table 2 section properties.
(6) C face and back must be natural unrepaired; if repaired, use stress level 2 (S-2).

Table 3.12 Effective Section Properties for Plywood
FACE PLIES OF DIFFERENT SPECIES GROUP FROM INNER PLIES
(INCLUDES ALL PRODUCT STANDARD GRADES EXCEPT THOSE NOTED IN TABLE 2.)

Nominal thickness (in.)	Approximate weight (psf)	t_s Effective thickness for shear (in.)	Stress applied parallel to face grain				Stress applied perpendicular to face grain			
			A Area (in.²/ft)	I Moment of inertia (in.⁴/ft)	KS Effective section modulus (in.³/ft)	lb/Q Rolling shear constant (in.²/ft)	A Area (in.²/ft)	I Moment of inertia (in.⁴/ft)	KS Effective section modulus (in.³/ft)	lb/Q Rolling shear constant (in.²/ft)
Unsanded Panels										
5/16-U	1.0	0.268	1.491	0.022	0.112	2.569	0.660	0.001	0.023	4.497
3/8-U	1.1	0.278	1.866	0.039	0.152	3.110	0.799	0.002	0.033	5.444
15/32- & 1/2-U	1.5	0.298	2.292	0.067	0.213	3.921	1.007	0.004	0.056	2.450
19/32- & 5/8-U	1.8	0.319	2.330	0.121	0.379	5.004	1.285	0.010	0.091	3.106
23/32- & 3/4-U	2.2	0.445	3.247	0.234	0.496	6.455	1.563	0.036	0.232	3.613
7/8-U	2.6	0.607	3.509	0.340	0.678	7.175	1.950	0.112	0.397	4.791
1-U	3.0	0.842	3.916	0.493	0.859	9.244	3.145	0.210	0.660	6.533
1-1/8-U	3.3	0.859	4.725	0.676	1.047	9.960	3.079	0.288	0.768	7.931

Table 3.12 Continued

Nominal thickness (in.)	Approximate weight (psf)	t_s Effective thickness for shear (in.)	Stress applied parallel to face grain				Stress applied perpendicular to face grain			
			A Area (in.²/ft)	I Moment of inertia (in.⁴/ft)	KS Effective section modulus (in.³/ft)	Ib/Q Rolling shear constant (in.²/ft)	A Area (in.²/ft)	I Moment of inertia (in.⁴/ft)	KS Effective section modulus (in.³/ft)	Ib/Q Rolling shear constant (in.²/ft)
Sanded Panels										
1/4-S	0.8	0.267	0.996	0.008	0.059	2.010	0.348	0.001	0.009	2.019
11/32-S	1.0	0.284	0.996	0.019	0.093	2.765	0.417	0.001	0.016	2.589
3/8-S	1.1	0.288	1.307	0.027	0.125	3.088	0.626	0.002	0.023	3.510
15/32-S	1.4	0.421	1.947	0.066	0.214	4.113	1.204	0.006	0.067	2.434
1/2-S	1.5	0.425	1.947	0.077	0.236	4.466	1.240	0.009	0.087	2.752
19/32-S	1.7	0.546	2.423	0.115	0.315	5.471	1.389	0.021	0.137	2.861
5/8-S	1.8	0.550	2.475	0.129	0.339	5.824	1.528	0.027	0.164	3.119
23/32-S	2.1	0.563	2.822	0.179	0.389	6.581	1.737	0.050	0.231	3.818
3/4-S	2.2	0.568	2.884	0.197	0.412	6.762	2.081	0.063	0.285	4.079
7/8-S	2.6	0.586	2.942	0.278	0.515	8.050	2.651	0.104	0.394	5.078
1-S	3.0	0.817	3.721	0.423	0.664	8.882	3.163	0.185	0.591	7.031
1-1/8-S	3.3	0.836	3.854	0.548	0.820	9.883	3.180	0.271	0.744	8.428
Touch-Sanded Panels										
1/2-T	1.5	0.342	2.698	0.083	0.271	4.252	1.159	0.006	0.061	2.746
19/32- & 5/8-T	1.8	0.408	2.354	0.123	0.327	5.346	1.555	0.016	0.135	3.220
23/32- & 3/4-T	2.2	0.439	2.715	0.193	0.398	6.589	1.622	0.032	0.219	3.635
1-1/8-T	3.3	0.839	4.548	0.633	0.977	11.258	4.067	0.272	0.743	8.535

From Plywood Design Specification by The American Plywood Association 1997.

Table 3.13 Effective Section Properties for Plywood (Structural I and Marine)
STRUCTURAL I AND MARINE

Nominal thickness (in.)	Approximate weight (psf)	t_s Effective thickness for shear (in.)	Stress applied parallel to face grain				Stress applied perpendicular to face grain			
			A Area (in.²/ft)	I Moment of inertia (in.⁴/ft)	KS Effective section modulus (in.³/ft)	Ib/Q Rolling shear constant (in.²/ft)	A Area (in.²/ft)	I Moment of inertia (in.⁴/ft)	KS Effective section modulus (in.³/ft)	Ib/Q Rolling shear constant (in.²/ft)
Unsanded Panels										
5/16-U	1.0	0.356	1.619	0.022	0.126	2.567	1.188	0.002	0.029	6.037
3/8-U	1.1	0.371	2.226	0.041	0.195	3.107	1.438	0.003	0.043	7.307
15/32- & 1/2-U	1.5	0.535	2.719	0.074	0.279	4.157	2.175	0.012	0.116	2.408
19/32- & 5/8-U	1.8	0.707	3.464	0.154	0.437	5.685	2.742	0.045	0.240	3.072
23/32- & 3/4-U	2.2	0.739	4.219	0.236	0.549	6.148	2.813	0.064	0.299	3.540
7/8-U	2.6	0.776	4.388	0.346	0.690	6.948	3.510	0.131	0.457	4.722
1-U	3.0	1.088	5.200	0.529	0.922	8.512	5.661	0.270	0.781	6.435
1-1/8-U	3.3	1.118	6.654	0.751	1.164	9.061	5.542	0.408	0.999	7.833

Table 3.13 Continued

Nominal thickness (in.)	Approximate weight (psf)	t_s Effective thickness for shear (in.)	Stress applied parallel to face grain				Stress applied perpendicular to face grain			
			A Area (in.²/ft)	I Moment of inertia (in.⁴/ft)	KS Effective section modulus (in.³/ft)	lb/Q Rolling shear constant (in.²/ft)	A Area (in.²/ft)	I Moment of inertia (in.⁴/ft)	KS Effective section modulus (in.³/ft)	lb/Q Rolling shear constant (in.²/ft)
Sanded Panels										
1/4-S	0.8	0.342	1.280	0.012	0.083	2.009	0.626	0.001	0.013	2.723
11/32-S	1.0	0.365	1.280	0.026	0.133	2.764	0.751	0.001	0.023	3.397
3/8-S	1.1	0.373	1.680	0.038	0.177	3.086	1.126	0.002	0.033	4.927
15/32-S	1.4	0.537	1.947	0.067	0.246	4.107	2.168	0.009	0.093	2.405
1/2-S	1.5	0.545	1.947	0.078	0.271	4.457	2.232	0.014	0.123	2.725
19/32-S	1.7	0.709	3.018	0.116	0.338	5.566	2.501	0.034	0.199	2.811
5/8-S	1.8	0.717	3.112	0.131	0.361	5.934	2.751	0.045	0.238	3.073
23/32-S	2.1	0.741	3.735	0.183	0.439	6.109	3.126	0.085	0.338	3.780
3/4-S	2.2	0.748	3.848	0.202	0.464	6.189	3.745	0.108	0.418	4.047
7/8-S	2.6	0.778	3.952	0.288	0.569	7.539	4.772	0.179	0.579	5.046
1-S	3.0	1.091	5.215	0.479	0.827	7.978	5.693	0.321	0.870	6.981
1-1/8-S	3.3	1.121	5.593	0.623	0.955	8.841	5.724	0.474	1.098	8.377
Touch-Sanded Panels										
1/2-T	1.5	0.543	2.698	0.084	0.282	4.511	2.486	0.020	0.162	2.720
19/32- & 5/8-T	1.8	0.707	3.127	0.124	0.349	5.500	2.799	0.050	0.259	3.183
23/32- & 3/4-T	2.2	0.739	4.059	0.201	0.469	6.592	3.625	0.078	0.350	3.596

From Plywood Design Specification by The American Plywood Association 1997.

Table 3.14 Allowable Stresses for Plywood (psi)*
ALLOWABLE STRESSES FOR PLYWOOD (psi) conforming to Voluntary Product Standard PS 1-95 for Construction and Industrial Plywood. Stresses are based on normal duration of load, and on common structural applications where panels are 24″ or greater in width. For other use conditions, see Section 3.3 for modifications.

Type of stress		Species group of face ply	Grade stress level[1]				
			S-1		S-2		S-3
			Wet	Dry	Wet	Dry	Dry only
EXTREME FIBER STRESS IN BENDING (F_b)	F_b & F_t	1	1430	2000	1190	1650	1650
TENSION IN PLANE OF PLIES (F_t) Face Grain Parallel or Perpendicular to Span (At 45° to Face Grain Use 1/6 F_t)		2, 3	980	1400	820	1200	1200
		4	940	1330	780	1110	1110
COMPRESSION IN PLANE OF PLIES	F_c	1	970	1640	900	1540	1540
		2	730	1200	680	1100	1100
Parallel to Perpendicular to Face Grain (At 45° to Face Grain Use 1/3 F_c)		3	610	1060	580	990	990
		4	610	1000	580	950	950
SHEAR THROUGH THE THICKNESS[3]	F_v	1	155	190	155	190	160
Parallel or Perpendicular to Face Grain (At 45° to Face Grain Use 2 F_v)		2, 3	120	140	120	140	120
		4	110	130	110	130	115
ROLLING SHEAR (IN THE PLANE OF PLIES)	F_s	Marine & Structural I	63	75	63	75	—
Parallel or Perpendicular to Face Grain (At 45° to Face Grain Use 1-1/3 F_s)		All Other[2]	44	53	44	53	48

Table 3.14 Continued

Type of stress	Species group of face ply	Grade stress level[1]				
		S-1		S-2		S-3
		Wet	Dry	Wet	Dry	Dry only
MODULUS OR RIGIDITY (OR SHEAR MODULUS)						
Shear in Plane Perpendicular to Plies (through the thickness) (At 45° to Face Grain Use 4G) G	1	70,000	90,000	70,000	90,000	82,000
	2	60,000	75,000	60,000	75,000	68,000
	3	50,000	60,000	50,000	60,000	55,000
	4	45,000	50,000	45,000	50,000	45,000
BEARING (ON FACE)						
Perpendicular to Plane of Plies $F_{c\perp}$	1	210	340	210	340	340
	2, 3	135	210	135	210	210
	4	105	160	105	160	160
MODULUS OF ELASTICITY IN BENDING IN PLANE OF PLIES						
	1	1,500,000	1,800,000	1,500,000	1,800,000	1,800,000
	2	1,300,000	1,500,000	1,300,000	1,500,000	1,500,000
E	3	1,100,000	1,200,000	1,000,000	1,200,000	1,200,000
For Grain Parallel or Perpendicular to Span	4	900,000	1,000,000	900,000	1,000,000	1,000,000

(1) See pages 12 and 13 for Guide.
To qualify for stress level S-1, gluelines must be exterior and veneer grades N, A, and C (natural, not repaired) are allowed in either face or back.
For stress level S-2, gluelines must be exterior and veneer grade B, C-Plugged and D are allowed on the face or back.
Stress level S-3 includes all panels with interior or intermediate (IMG) gluelines.
(2) Reduce stresses 25% for 3-layer (4- or 5-ply) panels over 5/8″ thick. Such layups are possible under PS 1-95 for APA RATED SHEATHING, APA RATED STURD-I-FLOOR, UNDERLAYMENT, C-C Plugged and C-D Plugged grades over 5/8″ through 3/4″ thick.
(3) Shear-through-the-thickness stresses for MARINE and SPECIAL EXTERIOR grades may be increased 33%. See Section 3.8.1 for conditions under which stresses for other grades may be increased.
* Stresses are based on normal duration of load, and on common structural applications where panels are 24 in. or greater in width.
From Plywood Design Specification by The American Plywood Association 1997.

3.4.9   Design Methodology and Design Equations

In selecting the sizes of plywood, joists, and stringers for given spans and loads, the following requirements must be considered.

1. The allowable working stresses for bending and for shear must not be exceeded.
2. The allowable limits for deflection must not be exceeded.
3. The sizes of the joists and stringers should be easily obtained in the local markets.

1. *Design for bending*:
   a. For single span:

   $$w_b = \frac{96F_b(KS)}{l_1^2}$$

   b. For two spans:

   $$w_b = \frac{96F_b(KS)}{l_1^2}$$

   c. For three or more spans:

   $$w_b = \frac{120F_b(KS)}{l_1^2}$$

   where

   w_b = uniform load for bending, lb/ft
   F_b' = adjusted allowable bending stress, psi
   KS = effective section modulus, in.^{3}/ft
   l_1 = span center to center of supports, in.

2. *Design for shear*:
   a. For a single span:

$$w_s = \frac{24F_s(Ib/Q)}{l_2}$$

b. For two spans:

$$w_s = \frac{19.2F_s(Ib/Q)}{l_2}$$

c. For three or more spans:

$$w_s = \frac{20F_s(Ib/Q)}{l_2}$$

where

w_s = uniform load for shear, lb/ft
F'_s = adjusted allowable rolling shear stress, psi
Ib/Q = rolling shear constant, in.^{2}/ft
l_2 = clear span, in. (center to center span minus support width)

3. *Design to satisfy deflection requirements*:
  a. Bending deflection:
     i. For a single span:

$$\Delta_b = \frac{wl_3^4}{921.6EI}$$

     ii. For two spans:

$$\Delta_b = \frac{wl_3^4}{2220EI}$$

     iii. For three and more spans:

$$\Delta_b = \frac{wl_3^4}{1743EI}$$

where

$$\Delta_b = \text{bending deflection, in.}$$

w = uniform load for bending, psf
E = modules of elasticity, psi
I = effective moment of inertia, in.^{4}/ft
l_3 = clear span + SW (support width factor)
SW = 0.25 in. for 2-in. nominal framing, and
 = 0.625″ for 4-in. nominal framing

b. Shear deflection: The shear deflection may be closely
    approximated for all span conditions by the following
    formula:

$$\Delta_s = \frac{wCt^2 l_2^2}{1270EI}$$

where

Δ_s = shear deflection, in.
w = uniform load, psf
C = constant, equal to 120 for panels with face
      grain parallel to supports and 60 for panels
      with face grain perpendicular to supports
t = nominal panel thickness, in.
E = modulus of elasticity, psi
I = effective moment of inertia, in.^{4}/ft

For cases when shear deflection is computed separately and
added to bending deflection to obtain the total deflection, E for
these bending-deflection equations should be increased by 10 percent. In this case the total deflection will be $\Delta = \Delta_b + \Delta_s$.

3.5 SLAB FORM DESIGN

3.5.1 Slab Form Components

As mentioned in the first two chapters, basic elements for an elevated slab formwork are sheathing, joists, stringers, shores,

wedges, and mudsills. Joists and stringers are designed as a system of main and secondary beams, as the joists rest on stringers. Design usually starts by first selecting a cross section for the elements and then calculating a corresponding span. An integer or modular value is usually selected for spacing.

3.5.2 Design Loads

Formwork should be designed to adequately sustain all the applied loads without failure or excessive deflection. ACI Committee 347-94 is considered the industry guide for estimating minimum and maximum loads applied on formwork. The following is a summary of the different load types and values.

Vertical Loads

Vertical load consists of dead load and live load. The weight of the formwork plus the weight of the freshly placed concrete is *dead load*. The *live load* includes the weight of workers, equipment, material, storage, impact, etc.

1. *Form's self-weight.* The self-weight of the form is usually assumed to be 5 lb/ft². Self-weight cannot be determined until the form is actually designed. After carrying out the design, the assumption of 5 lb/ft² needs to be checked. If the assumed value is far from the correct value, design should be repeated. This process is continued until one arrives at a reasonable difference between the two values. In most small and moderate spans, the 5 lb/ft² assumption holds well.
2. *Weight of concrete.* The weight of ordinary concrete is assumed to be 150 lb/ft³. Thus the load from the concrete slab on a square foot of decking will be

$$p = \frac{150 \times t}{12} = 12.5 \times t$$

where t is the thickness of slab in inches.

3. *Live load.* The American Concrete Institute (ACI = 347)
    specifies that different elements used to support vertical
    loads should be designed for a minimum live load of 50
    psf on the horizontal projection. When motorized cars are
    present, the minimum live load should be 75 psf.

The minimum design load for combined dead and live load
should be 100 psf, or 125 psf if motorized cars are present.

Horizontal Load

Braces and shores should be designed to resist all foreseeable horizontal loads such as seismic, wind, inclined support, starting and
stopping of equipment, and other such loads.

3.6 DESIGN STEPS

The steps in the design of forms to support concrete slabs include
the following:

1. Determine the total unit load on the floor decking, includ-
    ing the effect of impact, if any.
2. Select the type of floor decking, along with its net thick-
    ness.
3. Determine the safe spacing of floor joists, based on the
    strength or permissible deflection of the decking.
4. Select the floor joists, considering the load, type, size, and
    length of joists.
5. Select the type, size, and lengths of stringers, if required
    to support the joist.
6. Select the type, size, length, and safe spacing of shores,
    considering the load, the strength of the stringers, and
    the safe capacity of the shores.

Usually the most economical design of forms results when
the joists are spaced for the maximum safe span of the decking.

Likewise, reasonably large joists, which permit long spans, thus requiring fewer stringers, will be economical in the cost of materials and in the cost of labor for erecting and removing the forms. The use of reasonably large stringers will permit the shores to be spaced greater distances apart, subject to the safe capacities of the shores, thus requiring fewer shores and reducing the labor cost of erecting and removing them.

3.6.1   Size, Length, and Spacing of Joists

The selection of the size, length, and spacing of the joists will involve one of the following:

1. Given the total load on the decking, the spacing of the joists, and the size and grade of the joists, determine the maximum span for the joists.
2. Given the total load of the decking, the size and grade of the joists, determine the maximum spacing of the joists.
3. Given the total load on the decking, and the size and the span of the decking, determine the minimum size joists required.

For practical purposes, the selected span is rounded down to the next lower integer or modular value.

3.6.2   Stringers and Shores

The joist span selected will be based on the spacing of the stringers. We can follow the same procedure that is used for analyzing joists. Again, an integer or modular value is selected for stringer spacing.

After calculating the stringer spacing, the span of the stringer is checked against the capacity of the shore. The load on each shore is equal to the shore spacing multiplied by the load per unit foot of the stringer. The maximum shore spacing will be the lower of these two values (based on joists loading or shore spacing).

In calculating the design load of stringers, we do not consider the effect of joists on stringers. Instead, as a good approximation, we calculate load transmitted directly from sheathing to stringers. Hence, it is necessary to check for crushing at the point where the joists rest on stringers. Finally, shores are designed as columns (compression member). In checking the capacity of the different elements of the form, we use the same equations that will be used in the analysis of the wall form design. These equations are shown in Tables 3.15 and 3.16.

Table 3.15 Design Equations for Different Support Conditions

Type	One span	Two spans	Three spans
Bending moment (in.-lb)	$M = \dfrac{wl^2}{96}$	$M = \dfrac{wl^2}{96}$	$M = \dfrac{wl^2}{120}$
Shear (lb)	$V = \dfrac{wl}{24}$	$V = \dfrac{5wl}{96}$	$V = \dfrac{wl}{20}$
Deflection (in.)	$\Delta = \dfrac{5wl^4}{4608EI}$	$\Delta = \dfrac{wl^4}{2220EI}$	$\Delta = \dfrac{wl^4}{1740EI}$

Notation:
l = length of span (in.)
w = uniform load per foot of span (lb/ft)
E = modules of elasticity (psi)
I = moment of inertia (in.^{4})
Source: Reproduced from the 1998 edition of *Construction Methods and Management* by S. W. Nunnally, with the permission of the publisher, Prentice-Hall. Table 12-3, pp. 340–341.

EXAMPLE

Design formwork to support flat slab floor of 8-in. thickness and conventional density concrete. Sheathing will be plywood that has the following characteristics:

- Type: APA B-B plyform class I with species group of face ply = 2.
- Dry conditions.
- Thickness: 1⅛ in.

Table 3.16 Bending Moment, Shear, and Deflection Equations*

Design condition	Support conditions		
	One span	Two spans	Three or more spans
Bending			
Wood	$l = 4.0d\left(\dfrac{F_b b}{w}\right)^{1/2}$	$l = 4.0d\left(\dfrac{F_b b}{w}\right)^{1/2}$	$l = 4.46d\left(\dfrac{F_b b}{w}\right)^{1/2}$
	$l = 9.8\left(\dfrac{F_b S}{w}\right)^{1/2}$	$l = 9.8\left(\dfrac{F_b S}{w}\right)^{1/2}$	$l = 10.95\left(\dfrac{F_b S}{w}\right)^{1/2}$
Plywood	$l = 9.8\left(\dfrac{F_b KS}{w}\right)^{1/2}$	$l = 9.8\left(\dfrac{F_b KS}{w}\right)^{1/2}$	$l = 10.95\left(\dfrac{F_b KS}{w}\right)^{1/2}$
Shear			
Wood	$l = 16\dfrac{F_v A}{w} + 2d$	$l = 12.8\dfrac{F_v A}{w} + 2d$	$l = 13.3\dfrac{F_v A}{w} + 2d$
Plywood	$l = 24\dfrac{F_s Ib/Q}{w} + 2d$	$l = 19.2\dfrac{F_s Ib/Q}{w} + 2d$	$l = 20\dfrac{F_s Ib/Q}{w} + 2d$
Deflection	$l = 5.51\left(\dfrac{EI\Delta}{w}\right)^{1/4}$	$l = 6.86\left(\dfrac{EI\Delta}{w}\right)^{1/4}$	$l = 6.46\left(\dfrac{EI\Delta}{w}\right)^{1/4}$
If $\Delta = \dfrac{l}{180}$	$l = 1.72\left(\dfrac{EI}{w}\right)^{1/3}$	$l = 2.31\left(\dfrac{EI}{w}\right)^{1/3}$	$l = 2.13\left(\dfrac{EI}{w}\right)^{1/3}$

$$\text{If } \Delta = \frac{l}{240} \qquad l = 1.57\left(\frac{EI}{w}\right)^{1/3} \qquad l = 2.10\left(\frac{EI}{w}\right)^{1/3} \qquad l = 1.94\left(\frac{EI}{w}\right)^{1/3}$$

$$\text{If } \Delta = \frac{l}{360} \qquad l = 1.37\left(\frac{EI}{w}\right)^{1/3} \qquad l = 1.83\left(\frac{EI}{w}\right)^{1/3} \qquad l = 1.69\left(\frac{EI}{w}\right)^{1/3}$$

Notation:

l	=	length of span, center to center of supports (in.)
F_b	=	allowable unit stress in bending (psi)
F_bKS	=	plywood section capacity in bending (lb × in./ft)
F_c	=	allowable unit stress in compression parallel to grain (psi)
$F_{c\perp}$	=	allowable unit stress in compression perpendicular to grain (psi)
F_sIb/Q	=	plywood section capacity in rolling shear (lb/ft)
F_v	=	allowable unit stress in horizontal shear (psi)
f_c	=	actual unit stress in compression parallel to grain (psi)
$f_{c\perp}$	=	actual unit stress in compression perpendicular to grain (psi)
f_t	=	actual unit stress in tension (psi)
A	=	area of section (in.²)*
E	=	modulus of elasticity (psi)
I	=	moment of inertia (in.⁴)*
P	=	applied force (compression to tension) (lb)
S	=	section modulus (in.³)*
Δ	=	deflection (in.)
b	=	width of member (in.)
d	=	*depth of member (in.)*
w	=	uniform load per foot of span (lb/ft)

*For a rectangular member: $A = bd$, $S = bd^2/6$, $I = bd^3/12$.

Reprinted by permission of Prentice Hall

SOLUTION

Design loads:

- Dead load = weight of concrete + weight of the form-
  work.
  $$= \text{$^{8}/_{12}$ in. of concrete} \times 150 + 5.0 \text{ (assumed)}$$
  $$= 105 \text{ lb/ft}^2$$
- Live load = 75 lb/ft^2 (according to ACI = 347).
- Total vertical load = 75 + 105 = 180 lb/ft^2

Sheathing:

- Thickness = 1^1/$_8$-in. sanded panels.
- From Tables 3.12 to 3.14, we get the following properties:

  Grade Stress Level S-2.
  $A = 3.854$ in.^{2}
  $(KS) = 0.820$ in.^{3}/ft.
  $I = 0.548$ in.^{4}
  $lb/Q = 9.883$ in.^{2}/ft.
  $F_b = 1200$ psf
  $F_v = 140$ psf.
  $E = 1.2 \times 10^6$ psf

- Consider 1-ft strip. It carries a load of 180 × 1 =
  180 lb/ft.

From Table 3.15, using 3 or more spans:
Bending:

$$w_b = \frac{120 \, F_b \, (KS)}{l_1^2}$$

$$= 180 = \frac{120 \times 1200 \times 0.82}{l_1^2}$$

get

$$l_1 = 25.612 \text{ in.}$$

Shear:

$$w_s = 20F_v \frac{lb/Q}{l_2}$$

$$= 180 = \frac{20 \times 140 \times 9.833}{l_2} \rightarrow l_2 = 153.736 \text{ in.}$$

Deflection:

$$\Delta = \frac{wl_3^4}{1740EI}$$

$$< \frac{l_3}{360}$$

$$\frac{l_3}{360} = \frac{180 \times l_3^4}{1743 \times 1.2 \times 10^6 \times 0.548} \rightarrow l_3 = 26.055 \text{ in.}$$

Bending governs. Allowable span = 25.612 in., which is rounded down to 24 in. (2 ft). Hence, joist spacing = 2 ft.

Joists:

From Tables 3.6 to 3.8, we can get the following values for Redwood (select-structural):

- $F_b = 1350$ psi
- $F_v = 80$ psi
- $F_{c\perp} = 650$ psi
- $E = 1.4 \times 10^6$ psi
- Temperature factor $C_t = 1.0$.
- Load duration factor $C_D = 1.25$
- Shear stress factor $C_H = 2.0$ (no split case)

Choose 2 × 6 Redwood which has the following characteristics:

- $A = 8.25$ in.^{2}
- $S = 7.563$ in.^{3}
- $I = 20.8$ in.^{4}
- $d = 5.5$ in.
- $C_f = 1.20$

Load/ft of joist $= 180 \times 2 = 360$ lb/ft

Bending:

$$F'_b = F_b(C_f)\,(C_t)\,(C_D)$$

$$= 1350 \times 1.3 \times 1.0 \times 1.25 = 2193.75 \text{ lb/ft}$$

From Table 3.16

$$l = 10.95 \left(\frac{F'_b \times S}{w}\right)^{1/2}$$

$$= 10.95 \left(\frac{2193.75 \times 7.563}{360}\right)^{1/2} = 74.337 \text{ in.} = 6.195 \text{ ft}$$

Shear:

$$F'_v = F_v(C_H)\,(C_t)\,(C_D) = 80 \times 2 \times 1.0 \times 1.25 = 200 \text{ psi}$$

$$l = 13.5 \left(\frac{F'_v \times A}{w}\right) + 2 \times d$$

$$= 13.5 \times \frac{200 \times 8.25}{360} + 2 \times 5.5 = 72.875 \text{ in.}$$

Deflection:

$$E' = 1.4 \times 10^6 \text{ psi}$$

$$\Delta < \frac{l}{360}$$

$$l = 1.69 \left(\frac{EI}{w}\right)^{1/3}$$

$$= 1.69 \left(\frac{1.4 \times 10^6 \times 20.8}{360}\right)^{1/3} = 73.089 \text{ in.}$$

Shear governs. Span of joists $= 72.875$ in. $\cong 6$ ft (OK)

Stringers:

Load on a stringer is = $180 \times 6 = 1{,}080$ lb/ft
Use 2×8 (two stringers)

- $b = 1.5$ in.
- $d = 7.25$ in.
- $I = 47.63$ in.^{4}
- $A = 10.87$ in.^{2}
- $S = 13.14$ in.^{3}
- $E = 1.4 \times 10^6$ psi
- $C_f = 1.2$

Bending:

$$F'_b = 1350 \times 1.2 \times 1.0 \times 1.25 = 2025 \text{ psi}$$

$$l = 10.95 \times \sqrt{\frac{2025 \times 13.14 \times 2}{1080}} = 76.865 \text{ in.}$$

Shear:

$$l = 13.3 \times \left(\frac{F'_v \times A}{w}\right) + 2 \times d$$

$$= \frac{13.3 \times 200 \times (10.87 \times 2 \leftarrow \text{two stringers})}{1080}$$

$$+ 2 \times 7.5 \text{ in.} = 68.045 \text{ in.}$$

Note that we did not multiply the second term by 2 because, according to NDS, shear is checked at $d/2$ from the face of the support.

Deflection:

$$\left(\Delta < \frac{l}{360}\right)$$

$$l = 1.69 \times \left(\frac{1.4 \times 10^6 \times 47.63 \times 2}{1080}\right)^{1/3} = 84.157 \text{ in.}$$

Shear governs. Maximum span <68.045 in. (5.67 ft). Take stringer spacing = 5.5 ft.

Check for Crushing (joist on stringer):

Force transmitted from joist to stringer is equal to load of joist/ft multiplied by the span of the joist, force = 6 × 360 = 2160 lb.
Area through which this force is transmitted = 1.5 × 3 = 4.5 in.^{2}

Crushing stress = 2160/4.5 = 480 lb/in.^{2}
From Table 6.a, we get the following properties:

- $F_{c\perp} = 650$ lb/in.^{2}
- Bearing area factor = 1.25 ($b = 1.5$ in.)
- Temperature factor = 1.0

$F'_{c\perp} = F_{c\perp}(C_b)(C_t) = 650 \times 1.25 = 812.5$ psi since $480 < 812.5 \rightarrow$ safe in crushing.

Shore strength:

Required shore spacing = (stringer span)
Shore strength = span of stringer × load of stringer
$$= 5.5 \times 1080 = 5940 \text{ lb}$$

So we use shores whose strength is larger than 6000 lb.

4

Horizontal Formwork Systems:
Crane-Set Systems

Crane-set formwork systems are typically constructed and assembled as one large unit that can be moved horizontally or vertically from floor to floor by cranes. As a result, adequate crane services are required for handling these systems. This chapter deals with the following crane-set formwork systems: flying forms, column-mounted shoring system, and tunnel forms. These systems are characterized by their high initial cost and their rapid floor cycle time.

4.1 FLYING FORMWORK SYSTEM

Flying formwork is a relatively new formwork system that was developed to reduce labor cost associated with erecting and dismantling formwork. The name "flying formwork" is used because forms are flown from story to story by a crane. Flying form systems are best utilized for high-rise multistory buildings such as hotels and apartment buildings, where many reuses are needed.

4.1.1 Flying Formwork Components

Flying formwork is available in different forms that suit the particular needs of the project. The following components are found in most flying formwork systems available in North America and Europe. Figure 4.1 shows a model of flying truss system components.

Figure 4.1   Components of flying framework.

Sheathing Panels

Flying forms usually consist of large sheathing panels that are typically made of plywood or plyform. Plywood with standard size of
4×8 ft (1.22×2.44 m) allows fewer joints and produces a high
quality of concrete. The thickness of plywood used is a design
function. Figure 4.2 shows flying formwork system with a sheathing panel on the top.

Aluminum Joist "Nailers"

Sheathing panels are supported by aluminum "nailer-type" joists
(Figure 2.6). Each joist is a standard I beam with a wide top flange
that allows a wood nailer to be inserted to provide a wider nailing
surface for the sheathing panels. Other types of joists available are
symmetrically designed with wide top and bottom flanges that

Figure 4.2   Flying framework with sheathing panel on the top.

allow nailing strips on either side of the joist. Aluminum nailers are also shown at the top of the flying formwork system shown in Figure 4.1.

Aluminum Trusses

Sheathing panels and joists are supported by aluminum trusses. Aluminum trusses and joists are always used because of their light weight. However, steel trusses and joists are used for longer spans and heavier loads. Aluminum trusses are braced in pairs at the ground level to provide lateral stability in the direction perpendicular to the trusses (Figure 4.3).

Telescoping Extension Legs

Adjustable vertical telescoping extension legs are an integral part of the trusses; they are used to support the aluminum trusses and

Figure 4.3   Aluminum trusses braced in pairs.

to transfer the load vertically to the ground or to subsequent floors that have been already cast. Telescoping extension legs are made of square or circular hollow steel sections braced together by tabular steel struts (Figure 4.4). These legs can be adjusted up and down to achieve the exact level of the formwork. Extension legs are typically rested on wooden planks that distribute the loads to a larger area and also prevent the extension legs from sliding, particularly in the winter season.

For stripping, after the concrete has gained enough strength, the system can be lowered away from the slab by turning down the jacks. The truss mounted forms are then moved by crane from one casting position to the next.

4.1.2   Flying Formwork Cycle

Flying formwork is either assembled at the job site or preassembled in a local or regional yard facility and delivered to the job site.

Figure 4.4   Telescoping extension legs.

The flying formwork cycle can start from the ground floor if slabs-on-grade exist. The flying formwork cycle is described below. However, some minor technical details are different from one contractor to another. Figure 4.5 shows the four steps of flying formwork cycle.

Flying to a New Position

The flying tables are placed by crane into a new position in the bay between four or more adjacent columns or walls (step 1, Figure 4.5). The flying table is then lowered and placed on cribbing dollies (step 2, Figure 4.5). The adjustable extension legs are then extended to set the tables to the desired grades (step 3, Figure 4.5). Fillers are then placed over the columns to cover the space between columns and flying tables. Also, reinforcing steel, electri-

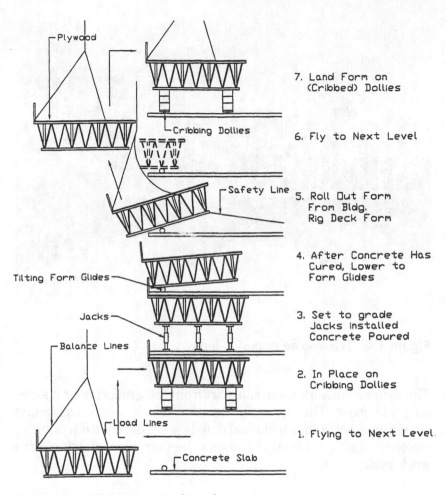

Figure 4.5 Flying formwork cycle.

cal and mechanical components, and any other services are installed. Concrete is then placed for slabs and parts of the columns.

Lowering and Stripping

After the concrete gains enough strength, the process of stripping
the flying tables begins. Stripping of flying formwork is carried out

by lowering the aluminum trusses and the attached deck (step 4, Figure 4.5). Lowering of the aluminum trusses can be performed by several hydraulic jacks. Hydraulic jacks are placed and fit under the bottom chord of the aluminum trusses to hold the table in place while the extension legs are retracted back. Hydraulic jacks are then used to lower the flying table onto its roll-out units.

Rolling Out

The lowered flying tables are then positioned onto roll-out units. The roll-out units are placed on the slab directly under the truss. Some roll-out units have wide cylinder flanges to facilitate fitting the truss bottom chord. Other roll-out units are made in a saddle shape that fits the bottom chord of the truss.

The flying table is then tilted and rolled out carefully by four construction workers (step 5, Figure 4.5).

Flying to a New Position

The table is carried by the crane, which is attached at four predetermined pick points. To prevent any swinging from the flying table, a safety line is normally attached between the lower chord of the aluminum truss and the concrete column. The table is then flown to its new position and the cycle is repeated (steps 6 and 7, Figure 4.5). It should be noted that occasionally the trusses only are carried out from floor to floor and the table is assembled in every floor or several floors. This is because of site limitation or because bay sizes and location are different from floor to floor. Figure 4.6 shows trusses carried out by crane without the intermediate nailers or sheets.

The total cycle time for the sequences described is approximately between 20 and 30 minutes, depending on the job conditions.

Figure 4.6 Aluminum trusses carried to next level.

4.1.3 Flying Formwork Usage and Benefits

Flying formwork has proved to be an efficient system in achieving a shorter construction cycle of initial fabrication, erection, stripping, and re-erection. Other visible benefits of flying formwork are as follows:

1. Fabrication of the flying formwork is normally performed on the ground, which yields higher productivity. Stripping flying formwork as one integral unit reduces the stripping costs to approximately 50 percent of the stripping costs for hand-set formwork systems such as conventional wood and conventional metal systems. Stripping of hand-set systems is performed by removing small pieces, which results in rather high labor costs.
2. Loads are transformed by telescoping extension legs located underneath the aluminum trusses and thus giving enough working space below the formwork to allow other

construction activities to be performed. In the traditional formwork system, several rows of shores are needed to provide support to the slab. These shores completely block any construction activity underneath the newly placed slabs for several days.

3. Costs of flying formwork are lower than for conventional horizontal formwork systems when 10 or more reuses are available. The high initial assembly cost is offset by a high number of reuses.

4. Lightweight aluminum trusses and joists allow average capacity construction cranes to handle the flying tables. Also, the lightweight aluminum joists can be placed on the aluminum trusses by one construction worker.

5. A shorter floor cycle can be achieved with use of the flying formwork system. A five-day construction cycle can be achieved for a medium-sized building of 100,000 ft^2 (9290 m^2). Reducing the floor cycle can shorten overall construction time, leading to substantial savings in overhead and financial costs.

6. A large-size flying table results in a smaller number of deck joints which produces high-quality smooth concrete.

7. Erecting and stripping the flying form as one large unit reduces the frequency of lifting work for the crane; this allows for crane time involvement with other construction work.

8. Fiberglass or steel pans used to form joist or waffle slabs can be placed on the flying tables and become an integral part of the flying table. These can be erected and stripped as one unit.

4.1.4   Flying Formwork Limitations

1. In windy weather conditions, large flying formwork panels are difficult to handle. In remote site conditions, the likely or higher chance of high wind may be a major factor in slowing the flying operation.

2. Flying formwork should not be used for flat slab with drop panels around the columns. Traditional formwork techniques should be used in this situation.
3. The building must have an open facade through which the flying tables can be passed. However, innovative construction methods developed collapsible flying tables that allow them to pass through restricted openings in the building's facade.

4.1.5. Modular and Design Factors for Selecting Flying Formwork

There are many factors that affect the selection of the flying formwork system for a concrete building. Some of these factors are related to economy, site condition, architectural, and structural considerations. The following section will focus on some of the dimensional considerations for selecting flying formwork. Architects and design engineers should be aware of these considerations so they can reduce building costs.

1. Standard modular flying tables are available for contractors to rent or purchase. Flying table width ranges from 15 to 30 ft (4.6 to 9.1 m), with the most economical width for flying tables being 22 ft (6.7 m). Standard aluminum truss height ranges from 4 to 6 ft (1.22 to 1.83 m). Total height with extension legs can reach 20 ft (6.10 m). As a result, flying formwork is limited to story height of 20 ft (6.10 m) maximum. It should be also noted that small flying tables are not economical.

    Large flying tables can reach a length of up to 120 ft (36.6 m) and a width of up to 50 ft (15.2 m). For long spans, two flying tables can be bolted together. For wide bay size, three aluminum trusses are needed to support wider tables. However, two trusses are sufficient for tables of up to 30 ft (9.1 m). Flying tables longer than 120 ft (36.6 m) are difficult to handle with average crane capacity.

2. Column size and location perpendicular to the flying table (i.e., the bay width) should be the same from floor to floor to avoid changing the formwork dimension by adding filler panels. It should be noted that a flying formwork system is not an economical alternative if the filler panels between flying tables exceeds 20 percent of the total area formed. The labor cost and time spent to add filler panels will negate the savings realized by using flying formwork.

3. Beam sizes and location should be the same from floor to floor on a modular building grid. Also, the depth of spandrel edge beams should be minimum, and cross beams should be avoided.

4.2   COLUMN-MOUNTED SHORING SYSTEMS

Column-mounted shoring system is the term used for formwork panels supported by an up-and-down adjustable bracket jack system attached to already-cast concrete bearing walls or columns. In contrast to traditional formwork systems, this formwork for slabs is supported by several levels of shores and reshores or the ground.

In multistory concrete buildings, the conventional construction method is to build formwork and place concrete for columns, then strip formwork for concrete columns after 12 to 24 hours. Erection of formwork and placing of concrete for slabs proceed after column forms are stripped. Though the newly placed columns have enough strength to support slab loads, they have a limited role in supporting the newly placed concrete slabs. The newly placed concrete slab is typically supported by several levels of shores and reshores. Those levels of shores delay or block the progress of any other construction activities underneath those concrete slabs. As a result, column-mounted shoring systems were developed to employ concrete columns to support formwork for concrete slabs and thus eliminate any need for shoring and reshoring that may ultimately reduce the overall construction schedule.

4.2.1 Components of Column-Mounted Shoring Systems

The system consists of two major components, a deck panel and a column- or wall-mounted bracket jack system. The deck panel consists of a plywood sheathing supported by a system of wood joists (headers), and a nailer-type open-web stringer that allows a 2×4 in. (50.8×101.6 mm) wood section to be inserted into the open web. A truss system supports both the joists and stringers. For smaller spans, a steel section can be used instead of the truss system. The truss system is supported by heavy steel I beams that run on all the sides of the deck panel. The I beam rests on the column-mounted jacks bolted in the concrete columns or bearing walls. Figure 4.7 shows a cross section of the column-mounted shoring system.

The second component of the column-mounted shoring system is the bracket jack system. The function of the bracket jack system is to support the deck panel. The weight of the freshly

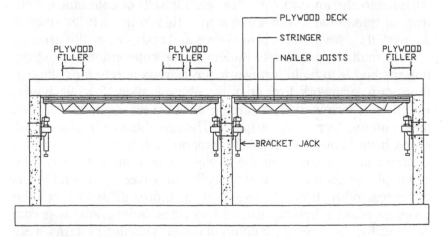

Figure 4.7 Components of column-mounted shoring systems. (Courtesy of Formwork Exchange Ltd.)

placed concrete and the dead weight of the deck panel are transferred from the deck through the bracket jack system and then to the concrete column or wall; thus no shores are required to support the deck panel. The bracket jack system has three major features:

1. A double steel roller at the top of the bracket jack where the deck's I beams rest, allowing the formwork deck panel to slide in and out with minimum effort. The steel roller unit can be adjusted and extended horizontally for up to approximately 10 in. (254 mm) by using an adjusting screw.
2. An adjustable screw to adjust the jack bracket and, consequently, the deck panel vertically. Maximum screw drop is 36 in. (914.4 mm) and standard screw drop is 18 in. (457.2 mm). There are two purposes for the vertical adjustment of the jack system: (a) to adjust the deck panel in its exact vertical position without needing to remove the bracket system, and (b) to lower the deck panel away from the slab during stripping.
3. A steel plate that contacts the concrete column or wall and is attached to the column or wall by two or four 1-in. (25.4-mm) through bolts.

Figure 4.8 shows the major components of the bracket jack system. It should be noted that the bracket jack weighs approximately 40 to 50 lbs and can be handled by one worker during installation and removal.

4.2.2   Column-Mounted Shoring System Cycle

Column-mounted shoring systems are either assembled at the job site or preassembled in a local or regional yard facility and delivered to the job site. Deck panels are flown from floor to floor by a crane in a manner similar to what is done in the flying truss system. Typically, the column-mounted shoring system cycle starts from the ground floor, whether or not slabs on grade exist.

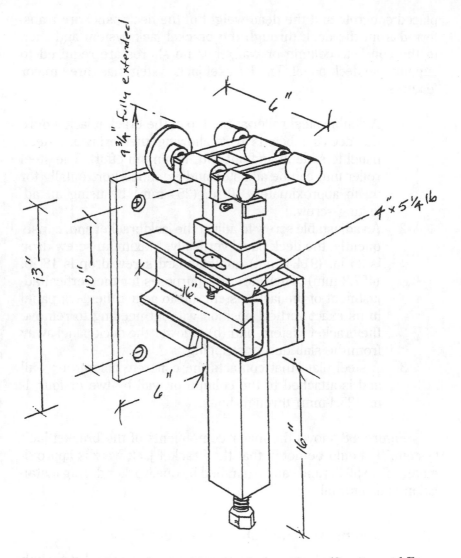

Figure 4.8 Components of bracket jack system. (Courtesy of Form-work Exchange Ltd.)

The column-mounted shoring system cycle is described below. However, some minor technical details are different from one manufacturer to another.

1. *Assembling the deck panel.* The deck panel can be assembled on the ground on the job site or preassembled in the factory. Assembling the deck panel at the job site starts by bolting the trusses to the flange of the I beam. The first truss is bolted first at one end of the I beam, then the center truss, and finally the truss at the other end of the I beam. Other trusses are then placed and bolted at regular intervals. Wood joists are then placed and attached to the trusses.

   Assembling the deck panel at the factory is achieved by attaching the trusses to only one I beam by one bolt so that the perpendicular trusses can be folded for shipment. At the job site, the trusses are unfolded and placed perpendicular to the I beam, and the other I beam is placed and attached to trusses. The remaining bolts are then fastened and tightened.

2. *Positioning of the deck panel.* The positioning starts by marking the elevation of the deck panel on the face of the column or the wall. The deck panel is then lifted by crane and positioned on the bracket jack system. The deck panel is lowered and adjusted to the previously marked elevation and then rested on the bracket jack system. Crane hooks and safety lines are then removed. After the deck is rested on the jack bracket system, installation of the fillers begins. Fillers are needed to fill the gaps over the concrete columns. Also, in cases in which the concrete column sizes are reduced at higher elevation, larger fillers are needed to compensate for this change in column sizes. In case the concrete columns are not aligned, a base plate is inserted between the jack bracket system and the concrete column or wall.

3. *Stripping of the deck panel.* After the concrete has been placed and has gained enough strength to support its own

weight and any additional loads, the deck panels are then stripped from the slab. The stripping process starts by lowering the jacks approximately 6 in. (152.4 mm) using the adjustable screws. The entire deck panel is then rolled out with use of a winch until the location of the first pair of crane hooks is outside the building line (exposed). Crane attachments and cables are then hooked and the deck panel continues to be pulled out until the second pair of crane hooks is exposed. Crane attachments and cables are hooked to the second pair of hooks and the deck panel is then carried by crane to the next floor.

In horizontal structures, deck panels can be rolled from one position to another on the rollers at the top of the jack bracket system.

4.2.3 Advantages, Usage, and Limitations of the Column-Mounted Shoring Systems

1. *Height-independent Systems.* The system can form slabs at higher elevation [15 ft (4.57 m) and above]. Traditional formwork systems are limited to story heights between 10 and 12 ft (3.05 and 3.66 m). Buildings with story height more than 12 ft (3.66 m) require several splices for shores; this slows the erection of formwork. In addition, construction workers are at risk because they are forming slabs at high elevation. Column-mounted shoring systems can be placed at higher elevation without the need for vertical posts. The fact that the system is height independent is of particular importance in the first few floors, where story height can reach 25 to 30 ft (7.6 to 9.1 m).
2. Because the system requires no shoring or reshoring, other construction activities are allowed to proceed simultaneously with the construction of concrete slabs. In conventional construction, shores that support the first slab are typically supported by mud sills that rest on the ground. Frequently, ground conditions are sloped,

rough, and/or irregular, causing concentration of loads on mud sills that might cause failure. Also, in the fall and spring seasons, day and night temperature vary substantially. Night temperatures cause a frozen ground condition, while high day temperatures in addition to excess water from construction activities make the frozen ground muddy, causing substantial settlement to mud sills. This may cause a serious safety problem. It should be noted that mud sills and/or slabs on grade are not required when column-mounted shoring systems are used.

3. Cost of the system is competitive with traditional systems if 8 to 10 reuses or more are available. The system requires a smaller crew size. Table 4.1 shows labor requirements for traditional systems and column-mounted shoring systems. The system can be assembled on the site at the ground level; this increases labor productivity. It should be noted that, on an average-size job, a four-person crew can strip, fly, and reset in approximately 20 minutes.

4. The system requires large capital investment (initial cost). However, if enough reuses are available, the final cost per square foot or square meter of contact area is competitive with traditional formwork systems.

5. The system requires adequate crane service in terms of adequate carrying capacity at maximum and minimum radii, adequate space around the building being con-

Table 4.1 Labor Requirements for Traditional Formwork Systems and Column-Mounted Shoring System

Traditional formwork	Column-mounted shorings
1 foreman	1 foreman
4 carpenters for slabs + 2 carpenters for shoring	3 carpenters
1 laborer	4 laborers
Crane operator may be needed	1 crane operator

structed, and nonexistence of power lines or any other obstructions that might limit crane movement and swing of the boom. Standard deck panel dimensions are 20 ft (6.1 m) wide by 40 ft (12.2 m) long and weigh approximately between 10 and 25 lb/ft² (48.8 and 122.1 kg/m²). Total weight of the average panel is between 10,000 and 12,000 lb (4540 and 5440 kg), which is within the capacity of most construction cranes. Larger panels can reach 30 ft (9.1 m) in width, 75 ft (22.9 m) in length, and 25 lb/ft² (122.1 kg/m²) in weight, requiring special cranes for handling.

6. The deck panel has an indefinite number of reuses as long as the plywood sheathing is of high quality and the panel is handled with care.

7. Electrical and plumbing connections can be installed on the ground on the deck panels. This results in substantial productivity improvement because there is no need for scaffolding or work performed at high elevations.

4.2.4 Modular and Design Factors Affecting Selection of Column-Mounted Shoring Systems

An important criterion when using a modularized formwork system is the maximization of the number of reuses. As a result, modification to the deck panels should be kept to a minimum. Modifications to deck panels are caused by changes in building design from floor to floor. The following are some modular limitations that should be considered to achieve maximum profitability of the column-mounted shoring system.

1. Optimum size of a column-mounted shoring systems deck is 16 to 20 ft (4.9 to 6.1 m) in width and 30 to 40 ft (9.1 to 12.2 m) in length. Maximum dimensions should not exceed 30 ft (9.1 m) in width and 70 ft (21.3 m) in length. If bay length is more than 70 ft (21.3 m), two deck panels can be bolted together.

2. Bay sizes should not vary from floor to floor, so as to avoid modification of the forms. In high-rise buildings where bay sizes changes are inevitable, bay widths should remain constant for at least 6 to 8 floors.
3. Columns and/or walls should be aligned to avoid adding shims. The addition of shims reduces the carrying capacity and adds more time for installing and removing of the jack bracket system.
4. Deep spandrel beams and cross beams between intermediate columns should be avoided. Spandrel beams should be 14 in. (355.6 mm) deep for economical use of this system. Deeper spandrel beams require extra shoring which increases the cost of the system and slows the progress of moving forms from floor to floor. Also, in the case of flat slabs, drop panels around the column should be avoided.

4.3 TUNNEL FORMWORK SYSTEM

The use of tunnel forms in North America is a relatively new process, but has been seen in several cities in Europe for a number of years. The growing popularity of the tunnel forms can be attributed to the high productivity that results in constructing buildings that have repetitive concrete rooms using tunnel forms. With the walls and slabs being cast monolithically, separate vertical and horizontal formwork systems are not required.

A tunnel form system is an ideal method to construct buildings with repetitive elements or rooms. Tunnel forms can be applied to both low- and high-rise buildings that require repetitive room design. Tunnel formwork has been used in the United States to construct resort hotels, condominiums, apartments, retirement homes, office buildings, hospitals, townhouses, and prisons. Once used on a particular project, they can be reused for continued return on investment.

4.3.1   System Description

Tunnel forms come basically in two different shapes: full and half.
Full-tunnel systems are all-steel formwork used for rooms that are
relatively square in shape. Half tunnel is an L-shaped all-steel formwork system that is set by crane, then another half-tunnel form
set separately adjacent to the previous half (Figure 4.9). The two
halves are then connected together to form an inverted U-shaped
tunnel form. The half-tunnel system allows for greater flexibility
in room sizes by simple addition or replacement of center fill panels. Half tunnels are simpler, lighter and faster in use than full
tunnels.

The tunnel formwork system consists of:

1. *Deck panel.* The thick steel skin used to form the ceiling
   and floor of each module
2. *Wall panel.* Also made of a thick steel skin, used to form
   the walls between two adjacent modules
3. *Waler and waler splices.* Stiffer deck and wall panels to
   minimize deflection due to concrete lateral pressure
4. *Diagonal strut assembly.* Used to provide additional support for the floor slab and keep walls and floors perpendicular

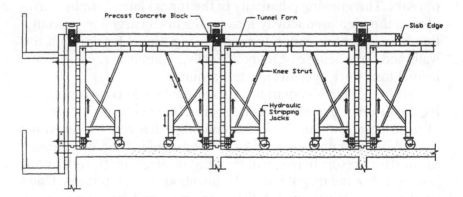

Figure 4.9   Tunnel formwork system. (Courtesy of Symons Corp.)

5. *Taper tie.* Wall tie between forms of two adjacent tunnel, keeps forms in place while concrete is being placed
6. *Wheel jack assembly.* Allows laborers to move tunnel forms over short distances, and also allows cranes to pull out a collapsed form after concrete has cured

4.3.2 Typical Work Cycle

Tunnel forms allow for swift erection of a building's skeleton. Each day a new room can be formed per tunnel for each tunnel form available. The following steps demonstrate the sequence of activities that cover the life cycle starting from the ground level.

Starter Slab and Wall

Construction usually begins with the casting of the ground slab to the desired configuration of the room. Included on the ground slab is a starter wall approximately 3 to 6 in. (76.2 to 152.4 mm) high, providing dimensional accuracy when setting the width of the forms (Figure 4.10).

First Cycle

Once the starter slab and wall have gained sufficient strength, the placement of steel reinforcement and the tunnel forms may proceed.

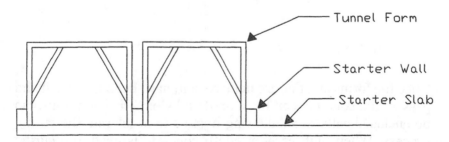

Figure 4.10 Starter slab and wall.

Steel Reinforcement Placement

The steel reinforcement may be placed before or after the tunnel forms are placed. When placed before the forms, the steel reinforcement for the walls is erected first, followed by placement of the tunnel form and finally the slab reinforcement. The second method uses preassembled sections of reinforcing steel for the walls and slabs. Once the forms are in place, the preassembled sections are lowered in place. Door and mechanical blockout, as well as electrical equipment, can be installed at this time.

Form Placement

The placement of the tunnel forms on the slab is done by crane. Once on top of the slab, the forms are rolled into place and set tightly against the starter wall. Hydraulic jacks attached to the form are then used to lift the forms to the proper elevation, usually 2 to 4 in. (50.8 to 101.6 mm) above the slab. Once these are in place, a precast concrete starter block is placed between the tops of the wall forms to provide proper wall thickness, set the slab depth and support the forms for the next starter wall. Form ties are attached at this time along with blockouts in the slab area for plumbing raceways, air-conditioning ducts, and other building services (Figure 4.10).

Concrete Pour

Once the forms and ties are in place, and have been inspected and passed, the concrete can be poured and vibrated. The rooms can be enclosed with tarps, allowing heaters to accelerate the curing process. When high early strength concrete is used, the curing process usually takes overnight.

Stripping

Once the concrete has gained sufficient strength, the tarps and heaters are removed, allowing the stripping to take place. The hydraulic jacks that were used initially to raise the forms to the correct elevation are reversed, pulling the forms away from the concrete slab above. Also attached to the tunnel forms is a diagonal strut system that allows easy stripping of the forms from the walls and deck with one simple movement (see Figure 4.11). The forms can now be rolled away from the walls and hoisted away by crane to be cleaned and oiled in preparation for the next location.

Second Cycle

Once the first cycle is complete, the forms are raised by crane once again and moved either to the next level or to a room adjacent to the one just completed.

A typical two-day cycle is shown in Figure 4.12. On the first day, two rooms have been poured while the third room has ended at a construction joint. If the third room is of the same size and

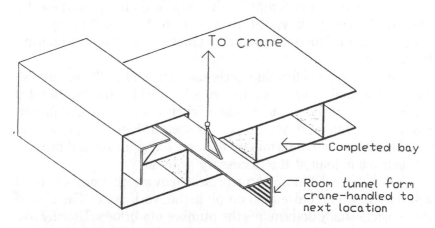

Figure 4.11 Raising the tunnel form.

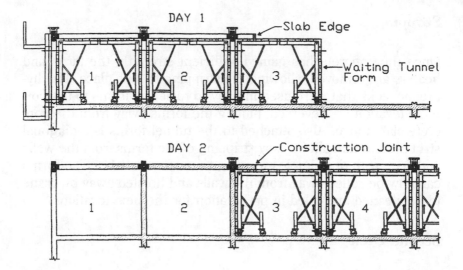

Figure 4.12 Typical two-way cycle.

dimensions, then a waiting tunnel form is typically used. On the second day, the pouring is started at the construction joint on the third room. The fourth room is also poured, while the fifth room is where the construction joint is now located.

Once one floor is complete, the cycle is then repeated for the next level starting above the first room to be cast. This type of cycle uses a third tunnel form that is always waiting for the pouring to catch up.

Another type of forming cycle uses an "end wall" tunnel wall after the second room. This system still provides the location for the construction joint, but is used when a corridor or different sized room is to be formed next. As seen in Figure 4.13, an end wall has been used to terminate the first day pouring and provide a construction joint at the necessary location.

On a typical tunnel form project, anywhere from 10 to 20 workers may be required to complete the work cycle. The size of the project usually determines the number of workers. Usually five to six rooms can be completed each day, using 10 to 20 workers, with five or six tunnel forms and one crane.

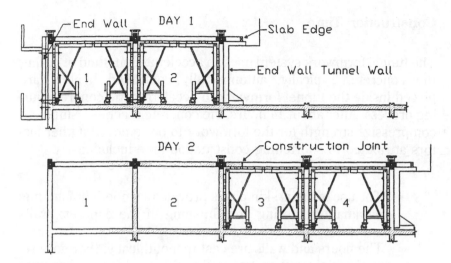

Figure 4.13 Typical two-day cycle with an end wall.

4.3.3 Advantages of Tunnel Formwork

Although relatively new, the tunnel form system is increasing in popularity due to its many advantages.

Cost

The initial cost for a formwork system may be deceiving when one is controlling the cost of the forming process. The initial cost for purchasing a set of five or six tunnel forms is approximately half of $1 million. Although this may be much higher than the cost of conventional forms, the all-steel formwork construction allows the number of reuses to be anywhere from 500 to a 1000. This allows the contractor to reuse the forms on similar projects, thus reducing the costs for each use of the forms. Once the forms begin to wear and deteriorate, rehabilitation of the forms can be accomplished for approximately $100,000. This increases the life and number of reuses of the forms.

Construction Time

The tunnel formwork system uses an accelerated method of curing the concrete. Typically, automatically controlled heaters are placed inside the tunnel formwork to accelerate the concrete curing process. After about 14 hours, the concrete reaches a sufficient compressive strength for the formwork to be removed. Other factors attributed to shortening construction time include:

- The use of metal skin plates provides a smooth finish, thus eliminating shoring and finishing of the concrete walls and slabs.
- The floors and walls are cast monolithically, therefore reducing time and cost as well as the need for two separate forming systems.
- Faster learning cycle: Once workers become familiar with the cycle, the time for each cycle will be reduced.

Labor Force

A major benefit of the tunnel forming system is that an experienced foreman can quickly turn relatively unskilled labor into skilled tunnel operators. This allows less expensive laborers to be used for other construction activities and can greatly reduce cost without sacrificing quality. This can be particularly beneficial in regions of high-cost labor. Also, since formwork transportation is handled by a crane, crew requirements are minimized.

Since the tunnel forms are prefabricated, the need to assemble and disassemble the forms between lifts is eliminated. The stripping of the forms is done by hydraulic jacks, thus reducing the number of workers needed for this task. The repetitive cycles allow the workers to become familiar with the system after 3 or 4 cycles, therefore increasing productivity as the job continues.

Flexibility

The tunnel formwork system has many characteristics that make it flexible in the construction site.

- The forms can be designed for different sized rooms. To use standard tunnel formwork, the wall should be at least 7.5 ft (2.29 m) high and should not be higher than 10 ft (3.04 m).
- The preassembled forms require no make-up area.
- The all-steel forms with fewer joints and ties provide an excellent concrete finish with little deflection.
- The tunnel forms can be reused 500 to 1000 times.
- Windows and doors can be easily moved from floor to floor. However, to achieve maximum productivity, opening sizes and location should be the same from floor to floor.

Quality

Because tunnel forms are preassembled in a factory-type setting, the uncertainty that often surrounds equipment fabricated on site is eliminated. The factory setting usually implements a quality assurance program to guard against imperfections or flaws in workmanship and materials. Also, because tunnel formwork is made of steel, the quality of the resulting concrete surface is superior and may require no additional finishing.

Safety

The tunnel forms have a number of safety features.

- Since they are preassembled in a factory setting, the system should have predictable strengths and quality assurance.

- The forms are not dismantled between lifts, and thus remain the same as when first used.
- Guard rails are provided at the slab edges.
- Supervised craning operations provide safety below.
- Workers positioning or stripping forms underneath are secure due to the all-steel and rigid tunnel form system.

4.3.4 Modular Consideration for Tunnel Formwork

Before the work cycle begins, the formwork designer must select the appropriate size and type of tunnel form. The overall dimensions of any tunnel form can be modified by simply adding fit-in sections (fillers). To increase the height, fit-in sections are added to the lower section of the walls, allowing the height to vary anywhere from 7.5 to 10 ft (2.29 to 3.04 m). The span of the tunnel form can also be varied by adding a fit-in section in the center of the deck, allowing the span to vary from 13 to 24 ft (3.96 to 7.32 m). As for the length, the standard dimension is 8 ft (2.44 m), but tunnels can be coupled to form varying lengths.

A majority of the reduction in construction cost and duration can be attributed to the standardization of building modules. The design begins by selecting a bay size that will be repeated time and time again. Because the design will dictate whether the tunnel forming method can be used, the construction process must be closely analyzed, organized, and coordinated from start to finish. This will ensure that construction can proceed with a minimum of labor, and over the shortest duration.

Consistency and simplicity are the easiest ways a designer can substantially reduce construction costs. Maintaining constant dimensions for the different structural elements unifies the project and allows for tunnel forming systems to be adapted. The repetitive room layout assures that the workers can increase productivity based on their learning

4.3.5 Limitations

The use of tunnel formwork is limited by specific design requirements. During the design phase of a project, the type of formwork should be determined. Without early consideration for a standard repetitive design and planning, the economic benefits of tunnel formwork cannot be achieved. There are also additional limitations that must be considered.

- High initial make-up cost.
- Requires extensive crane time.
- Requires high degree of engineering supervision.
- Takes workers a couple of cycles to become familiar with system.
- Used mainly for buildings with repetitive rooms.
- Large and different shaped rooms may be difficult.
- Site must be accessible since forms are large.
- To make the initial costs worthwhile, there should be at least 100 rooms.
- Can be used only on rooms with flat plate construction.

5
Selection Criteria for Horizontal Formwork System

This chapter provides a summary of the factors affecting the proper selection of horizontal formwork system. This chapter also presents a tabular comparative analysis of usage and limitations of each of the formwork systems presented in Chapters 3 and 4. An example of a formwork selection problem is also provided to explain how these tables can be used to accurately select the optimum formwork system for the job.

5.1 FACTORS AFFECTING HORIZONTAL FORMWORK SELECTION

Selecting the formwork system for cast-in-place reinforced concrete slabs is a critical decision that can affect cost, safety, quality, and speed of construction. Many factors must be considered for the proper selection of the formwork system. Among these are:

1. Factors related to building architectural and structural design, which include slab type and building shape and size
2. Factors related to project (job) specification, and schedule, which includes the speed of construction
3. Factors related to local conditions, which include area practices, weather conditions, and site characteristics
4. Factors related to the supporting organizations, which include available capital, hoisting equipment, home-office support, and availability of local or regional yard supporting facilities

An overview of all the factors affecting the selection of formwork systems is shown in Figure 5.1. The following sections briefly define the terminology and explain how these factors affect the selection of the horizontal formwork system.

5.1.1 Building Design: Slab Type

The construction cost of slabs is often more than half the cost of structural framing systems, except in extremely tall buildings. Therefore, selection of the slab formwork system deserves considerable attention to minimize cost.

The selection of a formwork system should be made on the basis of the selected floor system that satisfies the structural loading conditions. Floor slabs in concrete buildings are classified into two basic types, based on the load distribution applied on the slab:

1. Two-way slab, in which the rectangularity ratio (slab length/width) is between 1 and 2, and the slab load is transferred to the supporting beams in two directions. Two-way construction includes flat plate, flat slab, waffle slab, and two-way slabs supported by drop beams.
2. One-way slab, in which the rectangularity ratio (slab length/width) is more than 2, and the slab load is transferred to the supporting beams in one direction. One-way construction usually includes solid slabs on beams or walls, one-way joist (ribbed) slabs supported on beams or bearing walls.

Two-Way Flat Plate

A flat plate structural floor system consists of a concrete slab of constant thickness throughout, without beams or drop panels at the columns (see Figure 5.2a). Such slabs may be cantilevered at the exterior of the building to permit the use of exterior balconies.

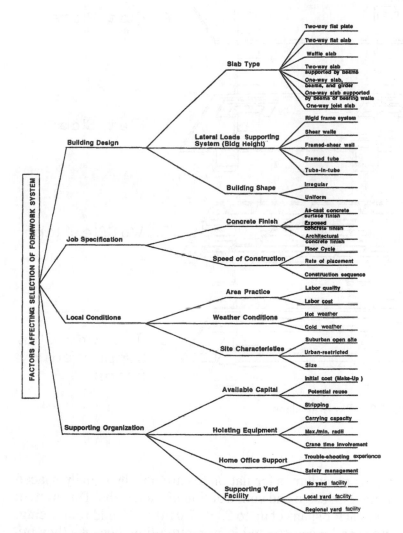

Figure 5.1 Factors affecting the selection of a formwork system.

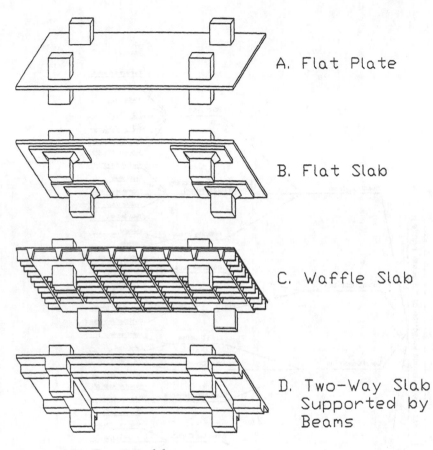

A. Flat Plate

B. Flat Slab

C. Waffle Slab

D. Two-Way Slab
   Supported by
   Beams

Figure 5.2   Two-way slabs.

The supporting columns for flat plates are usually equally spaced to facilitate the design and construction of such slabs. This system is economical for spans of up to 23 ft (7.0 m) with mild reinforcing. Flat plates can be constructed in minimum time because they utilize the simplest possible formwork. Flat plates have been used successfully in multistory motel, hotel, hospital, and apartment buildings.

Two-Way Flat Slab

A flat slab structural system consists of a constant thickness of concrete slab with drop panels at the columns locations (see Figure 5.2b). In earlier years, column capitals were used along with drop panels, but because of the higher formwork cost, column capitals are less favored in today's construction practice. Flat slabs are used to resist heavier loads and longer spans than flat plates. Generally, the system is most suitable for square or nearly square panels.

Waffle Slab

Waffle slab construction is shown in Figure 5.2c. It consists of rows of concrete joists at right angles to solid heads at the columns. Waffle slabs can be used for spans up to about 50 ft (15.2 m), and they are used to obtain an attractive ceiling.

Two-Way Slab Supported by Beams

This system consists of a solid slab designed to span in two directions, to either concrete beams or walls (see Figure 5.2d). The primary advantage of the system is the saving in reinforcing steel and slab section as a result of being able to take advantage of two-way action. Formwork for the two-way system is complicated and usually outweighs the cost advantages associated with the saving in reinforcing steel and slab thickness.

One-Way Slab, Beam, and Girder

This system consists of a solid slab, spanning to concrete beams which are uniformly spaced. The beams, in turn, are supported by girders at right angles to the beam to carry loads into the columns (see Figure 5.3a). This system generally provides the opportunity to span longer distances than two-way by designing deeper beams and girders.

One-Way Slab Supported by Beams or Bearing Walls

This system is a modification of the slab, beam, and girder system. It eliminates the secondary beams (see Figure 5.3b). Reinforcing steel is relatively simple, and existence of openings is generally not a critical concern.

One-Way Joist (Ribbed) Slab

One-way joist slabs are a monolithic combination of uniformly spaced beams or joists and a thin cast-in-place slab to form an integral unit. When the joists are parallel, it is referred to as one-way joist construction (see Figure 5.3c). Joists are very attractive to architectural layout and mechanical support systems.

5.1.2   Building Shape

Special buildings such as industrial buildings and power plants usually have extensive electrical and mechanical requirements which do not lend themselves to any sophisticated formwork system. As a result, they should be constructed using the traditional formwork method.

Some of the factors that enable the contractor to decide whether to use a formwork system or a traditional forming method are:

1. Variation of column and wall location
2. Variation of beam depth and location
3. Variation of story height
4. Existence of blockouts and openings for windows and doors
5. Extensive HVAC requirements

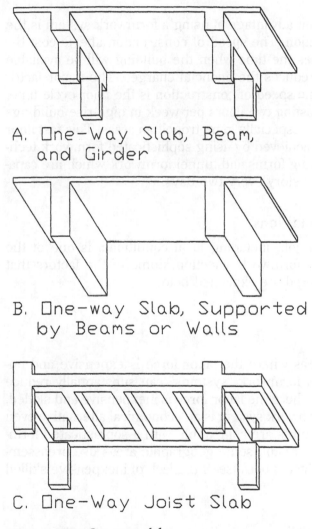

A. One-Way Slab, Beam, and Girder

B. One-way Slab, Supported by Beams or Walls

C. One-Way Joist Slab

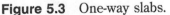

Figure 5.3 One-way slabs.

5.1.3 Job Specification
Speed of Construction

The most important advantage of using a formwork system is the speed of construction. The speed of construction affects cost because it determines the time when the building will be available for use and also reduces the financial charges. The major factor that determines the speed of construction is the floor cycle time. In recent years, casting two floors per week in high-rise buildings has been achieved, especially in metropolitan areas. This fast floor cycle can only be achieved by using sophisticated formwork techniques such as flying forms and tunnel formwork which are capable of forming one story every two days.

5.1.4 Local Conditions
The nature of the job, including local conditions, is one of the primary factors in formwork selection. Some of the factors that should be considered are explained below.

Area Practice

In geographic areas where the labor force is expensive and unskilled, the use of formwork "systems" can substantially reduce the cost. In areas where the labor force is inexpensive and skilled, a conventional formwork system is an economical alternative even if the building features are compatible with a sophisticated formwork system. As a result, some geographic areas use preassembled formwork systems because of the lack of inexpensive skilled labor force.

Site Characteristics

The building site itself may influence the selection of a suitable forming system, because of site limitations and accessibility for construction operations. The feasibility of using flying forms, for instance, is influenced by site characteristics, which include:

1. Accessibility to the site.
2. Availability of a fabrication area.
3. Surrounding area restrictions such as property lines, adjacent buildings, power lines, and busy streets. In open and unrestricted suburban sites, all forming systems are practical and some other considerations should be evaluated to determine the most efficient and cost-effective system. In downtown restricted sites, the only possible system may be ganged units that can be transferred from floor to floor.

5.1.5 Supporting Organization

Most of the crane-set formwork systems (i.e., flying form, column-mounted shoring system, and tunnel), require high initial investment and intensive crane involvement. The major resource requirements that should be carefully evaluated when deciding upon a forming system are discussed below.

Available Capital (Cost)

The cost of concrete formwork is influenced by three factors:

1. Initial cost or fabrication cost, which includes the cost of transportation, materials, assembly, and erection.
2. Potential reuse, which decreases the final total cost per square foot (or per square meter) of contact area. The data in Table 5.1 indicates that the maximum economy can be achieved by maximizing the number of reuses.
3. Stripping cost, which also includes the cost of cleaning and repair. This item tends to remain constant for each reuse up to a certain point, at which the total cost of repairing and cleaning start rising rapidly.

Table 5.1 Effect of Reuse on Concrete Formwork Cost Based on One Use Equal to 1.00

Number of uses	Cost per square foot of contact area	Cost per square meter of contact area
One	1.00	10.76
Two	0.62	6.67
Three	0.5	5.38
Four	0.44	4.74
Five	0.4	4.31
Six	0.37	3.98
Seven	0.36	3.88
Eight	0.35	3.77
Nine	0.33	3.55
Ten	0.32	3.44

In deciding to use a specific formwork system, the initial cost should be evaluated versus the available capital allocated for formwork cost. Some formwork systems tend to have a high initial cost, but through repetitive reuse, they become economical. For example, slipforms have a high initial cost, but the average potential reuse (usually over 100) reduces the final cost per square foot (or per square meter) of contact area of this alternative. In the case of rented formwork systems, the period of time in which the formwork is in use has a great effect on the cost of formwork.

Hoisting Equipment (Cranes)

Some formwork systems require special handling techniques, which can include a good crane service. The flying truss system is a good example of crane influence on the selected system. The size of the flying modules may be limited by the crane carrying capacity and its maximum and minimum lift radii.

Supporting Yard Facility

The feasibility of using prefabricated forms such as flying formwork is largely influenced by the availability of a local or central (regional) yard facility. When a local or central yard facility is available, the standard formwork elements can be manufactured and assembled under efficient working conditions. However, the cost of transporting form sections to the site may influence the economy of the selected system.

5.2   CHOOSING THE PROPER FORMWORK SYSTEM USING TABLES

Table 5.2 shows the relationship between the factors affecting the selection of formwork systems and the different forming systems available for horizontal and vertical concrete work. The user must first list all the known major components of their project and then compare them to the characteristics listed in the table under each forming system. The best formwork system can then be identified when the project features agree with most of the characteristics of particular system. The following example shows how Table 5.2 can be used to identify the best formwork system for horizontal concrete work.

5.2.1   Example Project

A 14-story concrete building is to be located at 1601 Pennsylvania Avenue, Washington, D.C. Building size is approximately 22,500 ft^2 (2090 m^2) per floor. Floor slabs are 8-in. (203.2-mm) flat slab with drop panels at every column. Column sizes and locations vary due to the existence of a three-story high entrance, free from columns. Story heights vary from 14.5 in. (368.3 mm) for the first three floors to 10.5 in. (266.7 mm) for the remaining eleven stories. There are no cantilevered balconies, and the slab on grade will not be in place before forming operations start. The building is located in a highly restricted downtown area.

Existing buildings and traffic limit the movement of equipment on all sides of the building. The area has a highly qualified labor force and high hourly labor costs.

Table 5.2   Factors Affecting the Selection of Horizontal Forming Systems

Influence factor		Conventional wood system	Conventional metal system	Flying truss system	Column-mounted shoring system	Tunnel form
Slab type (Slab type / Lateral support)	Slab type	All slab systems Most suited for two-way slab supported by beams or one-way slab, beam, girder		Two-way: flat plate and flat slab One-way: slab supported by beams or walls and joist slab (standard or skip-joist)		One-way slab supported by walls Less than 12-in. thickness
	Compatibility with lateral load supporting system	Compatible with all lateral load supporting systems		Generally not suitable for framed tube and tube in tube because of the close distance between columns which characterize the tube systems		Bearing wall
Building shape	Horizontal uniformity / irregularity	Can handle variations in beam size and location Can handle variations in cantilever shape, size, and location Avoid cross beams		Beams should be of the same size and location or within 20% difference from floor to floor Cantilever should be of the same size and location or within 20% difference from floor to floor		Beams should be of the same size and location Cantilever balconies should be of the same size and location
	Vertical uniformity / irregularity	Can handle variations of column/wall size and location. Can handle variations in story height within one floor or from floor to floor		Column/wall should be of the same size and location or within 20% difference from floor to floor Can handle limited variation (20%) in story height		Walls should be of the same size, location, and height from floor to floor
	High stories (higher than 14 ft)	Not suitable for high stories	More suitable for high stories (light alum. wt.)	Limited by truss depth (up to 20 ft)	Height independent system	Limited height system (between 7.5 and 10 ft)
Miscellaneous	Openings	System can handle variation in opening size and location		Can handle limited variation (20%) in opening size and location		
	Slopes and cambers	Slopes and cambers can be accommodated at additional cost	System must be designed to accommodate slopes and cambers	Slopes and camber must be identical from floor to floor		
	HVAC	Can accommodate extensive HVAC	Cannot accommodate extensive HVAC	HVAC should be minimal and identical from floor to floor		
	Dimension Limitations	Used for small building size (less than 100,000 ft²)	Used for medium building size (between 100,000 and 200,000 ft²)	Used for large size buildings (more than 200,000 ft²)		

Building design

Table 5.2   Continued

Influence factor		Formwork system	Conventional wood system	Conventional metal system	Flying truss system	Column-mounted shoring system	Tunnel form
Job specification	Speed of construction	Floor cycle (number of floors /day)	Typically floor every 5 working days Faster cycle can be accommodated at additional cost (increasing number of stories to be shored and reshored)		Floor every 3–4 day		Floor every 1–2 days
		Rate of placement	Generally, not a major factor in horizontal concrete work, average rate between 25–30 cycles/hours				
		Construction sequence	Pouring columns, then beams and slabs Slab on grade is not necessarily to be in place, but cost can be reduced if slab on grade is in place		Pouring columns, then beams and slabs Slab on grade should be in place for the form to be used in first floor	Pouring columns, then beams and slabs Slab on grade is not necessarily to be in place	Pouring slabs and walls together 3-in. starter wall is necessarily
Local conditions	Site characteristics	Area pratice	Work best in areas of high-quality, low-cost labor force	Work best in areas of high-quality, low-cost labor force	Generally, work in areas of high-cost, low-quality labor force		
		Available storage make-up area	Require minimum storage and make-up area	Small make-up area is required if system is panelized	Site must have adequate storage and make-up area		System is preassembled Minimum storage area is required.
		Access to site	Generally, not a factor for hand-set systems		Job must be acessible for large form units May not be a factor if forms are assembled on site		Job must be accessible for large preassembled steel forms

Table 5.2 Continued

Formwork system / Influence factor	Conventional wood system	Conventional metal system	Flying truss system	Column-mounted shoring system	Tunnel form
Available capital — Initial "make-up" cost	Average cost range from $1 to $3 per square foot of contact area		Average cost range from $10 to $15 per square foot of contact area		Average cost ranges from $20 to $50 per square foot of contact area
Potential reuse	Up to 15 reuses	Up to 20 reuses	Minimum of 12–15 reuses should be available	Minimum of 15–20 reuses should be available	Minimum of 50 reuses should be available
Stripping cost	High stripping cost (approximately one-third of the makeup cost)		Low stripping cost (approximately one-half of the hand-set systems)		Low stripping cost
Average labor productivity	12 ft² per contact area	18 ft² per contact area	36 ft² per contact area	45 ft² per contact area	50 ft² per contact area
Hoisting equipment — Availability of crane and crane time	Can be hand-set Less expensive if crane is available	Can be hand-set Crane is necessary If made into panels	Adequate hoisting equipment must be available		
Supporting organizations — Adjacent building traffic and other obstructions	Generally, not a factor	Generally, not a factor May be a factor if system panelized	A major factor, there must be open space at least 1.5 the length of the large panel from the face of the building		
Adequacy of crane carrying capacity	Generally, not a factor		Crane should have adequate carrying capacity at maximum and minimum radii.		

Table 5.2 Continued

Influence factor	Formwork system	Conventional wood system	Conventional metal system	Flying truss system	Column-mounted shoring system	Tunnel form
Supporting organizations / Home office support	Safety management	Normal safety precautions are required		Special safety in dealing with crane operation		
	Supervision "Troubleshooting experience"	Min. engineering supervision	Requires skilled crew / Requires moderate engineering supervision	Requires high engineering supervision		Requires high degree of engineering supervision
	Supporting yard facility	Min. yard is required	Requires local yard if made into panels	Requires enough supporting local or regional yard facility		System should be available in close by area / No yard is required

5.2.2   Use of Formwork Tabular Comparative Analysis

The fact that a tunnel form is used for only one-way slabs supported by a wall makes this system (tunnel form) an inappropriate choice. It should therefore be eliminated. Also, the potential number of reuses (14) cannot justify the use of tunnel forms which require at least fifty reuses. Flying truss and column mounted shoring systems are also eliminated because of the restricted site characteristics in downtown Washington D.C. (Pennsylvania Avenue). Crane movement is limited even though adequate crane service is available. Also, the irregular column spacing strongly suggests the elimination of these systems. As a result the choice is narrowed to either conventional wood or aluminum systems. A review of Table 5.2 reveals that a conventional aluminum system is a more appropriate selection than a conventional wood system for the following reasons:

1.   The building size is 315,000 ft^2 (29,300 m^2), which is more appropriate for the aluminum system (look at building shape "dimension limitations").
2.   The story height in the first three floors is 14.5 ft (4.42 m) (look at height stories).
3.   The area is characterized by high quality and expensive labor force (look at area practice).

It should be noted that the conventional wood system can be used, but the conventional aluminum system is more appropriate.

6

Vertical Formwork Systems: Crane-Dependent Systems

6.1 INTRODUCTION TO VERTICAL FORMWORK SYSTEMS

Formwork development has paralleled the growth of concrete construction throughout the twentieth century. As concrete has come of age and been assigned increasingly significant structural tasks, form manufacturers have had to keep pace. Form designers and builders are becoming increasingly aware of the need to keep abreast of technological advancements in other materials fields in order to develop creative innovations that are required to maintain quality and economy in the face of new formwork challenges.

Formwork was once built in place, used once, and subsequently wrecked. The trend today, however, is toward increasing prefabrication, assembly in large units, erection by mechanical means, and continuing reuse of forms. These developments are in keeping with the increasing mechanization of production in other fields.

Vertical formwork systems are those used to form the vertical supporting elements of the structure—columns, core walls, and shear walls. The functions of the vertical supporting systems are to transfer the floor loads to the foundation and to resist the lateral wind and earthquake loads. Consequently, the construction of vertical structural elements precedes flat horizontal work. Typical vertical formwork systems utilized in construction include conventional formwork, ganged forms, jump forms, slipforms, and self-raising forms.

Formwork systems for vertical concrete work can be classified into two main categories, namely, crane-dependent systems and crane-independent systems. Gang formwork and jump form are classified under crane-dependent systems. On the other hand, slipform and self-raising formwork are classified as crane independent systems in which formwork panels are moved vertically by other vertical transportation mechanisms. This chapter focuses primarily on crane-dependent formwork systems and their application and limitations.

The conventional wall system is the only hand-set system. The other four formwork systems are made of prefabricated modular panels before they can be transported by cranes or any other vertical transportation system.

6.2 CONVENTIONAL WALL/COLUMNS FORMING SYSTEMS

This all-wood forming system consists of sheathing made of plywood or lumber that retains concrete until it hardens or reaches adequate strength. This system is also known as job-built wood system. The sheathing is supported by vertical wood studs. The studs are supported by horizontal wales which also align the forms. Single or double horizontal wales are used to support the studs (Figure 6.1). However, double wales are preferred to avoid drilling through single wales, which reduces its load-carrying capacity. Ties are drilled through wales (single wale) or inserted between them (double wale) to resist the lateral pressure of plastic (wet) concrete. An inclined bracing system is used to resist construction and wind loading that formwork is subject to.

6.2.1 System Components and Construction Sequence

Components of the conventional wall system are similar to conventional wood system components for slabs but have different names. Joists become studs and stringers become wales. Also, the two systems are similar in that they are built in situ and stripped piece by piece.

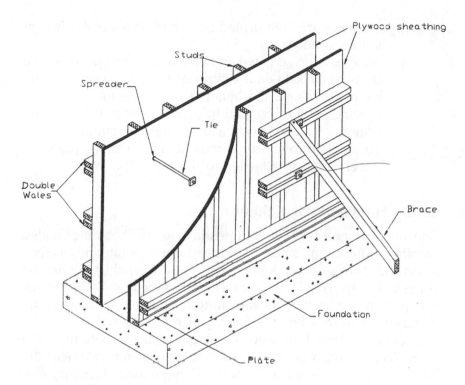

Fig. 6.1 All-wood conventional wall-forming systems.

Erection sequence for all wood conventional wood system is as follows:

1. Erection of wall form starts by attaching the first side of the plywood to the concrete footing wood sill (shoe) by anchors or hardened nails. The plywood is erected with the longer direction parallel to the length of the wall.
2. Studs are then erected and temporarily supported by wood bracing [usually 1×6 in. (25.4×152.4 mm) brace]. Reinforcing steel, opening boxouts, and other electrical or mechanical systems are installed before the second side of the wall is erected. The plywood is then nailed to the studs and the other wall side (plywood) is then erected.

3. Tie holes are then drilled from both sides of the wall at proper locations.
4. Wales are then erected and attached to the outside of the studs by nails. In double-wale systems, each wale should be located above and below the tie location.
5. Bracing is then installed to support horizontal loads resulting from wind loads and concrete vibration.
6. To facilitate concrete placement and finishing, scaffolds are erected and attached to the top of the wall.

6.2.2   Formwork for Columns

Conventional formwork for columns is made of sheathings nailed together to form rigid sides. Typically, formwork for concrete columns has four sides. Column form sides are held together by yokes or clamps (see Figure 6.2). Another function of these yokes is to prevent the buckling of sheathing resulting from the horizontal lateral pressure when the fresh concrete is placed.

Concrete lateral pressure is greater near the bottom of the form. As a result, yokes are spaced at smaller intervals near the bottom than near the top of the form. Column form sides may also be tied by straps or steel angles. In order to prevent breaking of the corners or edges, it is common practice to add a triangular fillet to the form along the edges of the columns. This practice also facilitates stripping of column forms.

Columns can take several shapes: round, rectangular, L-shaped, or various irregular shaped cross sections. Irregular shapes are frequently formed by attaching special inserts inside square or rectangular forms.

6.2.3   Erection of Column Forms

Erection starts by marking a template on the floor slab or footing to accurately locate the column floor. Erection sequence is somewhat similar to wall forms; however, methods vary depending on the available lifting equipment and whether reinforcing cages and forms are built in place or not.

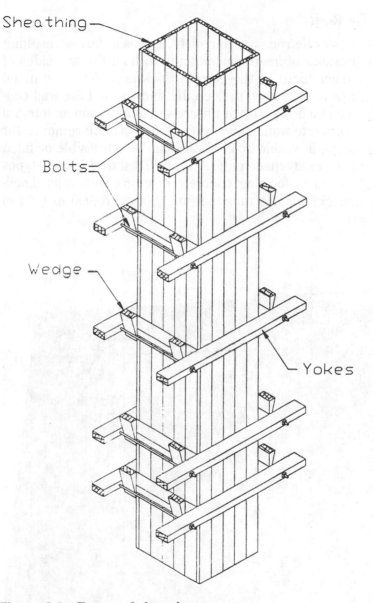

Figure 6.2 Formwork for columns.

6.2.4   Tie Rods

The functions of tie rods are to resist the tensile forces resulting from the pressure of fresh concrete and to hold the two sides of wall form (sheathing) at the correct thickness. Wood or metal spreaders can also be used to keep the thickness of the wall constant. Ties can be broken off or unscrewed and remain an integral part of the concrete wall. Other types of ties may be removed for reuse, resulting in visible holes. Holes can be left visible or filled with mortar or ready-made plugs. Figure 6.3 shows several types of tie rods used in forming concrete columns and walls. Load-carrying capacity for ties ranges from 1,000 to 70,000 lb (450 to 31,750 kg).

Figure 6.3   Wall form ties.

6.2.5 Construction Practices

1. It is good practice to minimize cutting formwork material to suit the wall size. Plywood, studs, and wales may be extended beyond the size of the wall and concreting is stopped at the appropriate size. For example, the drawings may call for a wall to be 11 ft (3.35 m) high. Plywood and studs can be extended to 12 ft (3.66 m) high and concreting can be stopped when it reaches 11 ft (3.35 m) high.
2. In long studs or wales where more than one piece is needed, joists between different pieces should be staggered to avoid creating a plane of weakness.
3. When placing concrete for tall columns, it is recommended to have pockets or windows at mid-height or other intervals to facilitate placing and vibrating the concrete.

6.2.6 Economy of Conventional Wall Formwork

Conventional wall formwork systems are economical when a limited number of reuses are expected and wall or column configurations are not repetitive. The expected number of reuses for conventional job-built forms is three to four times, depending on the quality of wood, connecting hardware, and handling of the wood during erection and stripping.

The limitations of using conventional wood systems for concrete walls are similar to those of conventional slab forms, namely, high labor costs and materials waste.

6.3 GANGED FORMING SYSTEMS

Ganged forms are large wall form units that are made of panels joined together with special hardware and braced with strongbacks or special steel or aluminum frames. Gang forms can be made on the site, rented, or purchased from formwork manufacturers. The advantages of manufactured forms over site made is that they are precise in dimension and can be reused a larger number of times.

6.3.1   Sizes and Materials

Sizes of gang forms vary substantially from smaller units that are handled manually, to much larger units that are handled and raised by cranes. Smaller gang forms are typically 2 × 8 ft (0.61 × 2.44 m) and 4 × 8 ft (1.22 × 2.44 m), and weigh between 50 and 100 lb (23 and 45 kg). Larger gang forms are limited by crane carrying capacity and can reach 30 × 50 ft (9.1 × 15.2 m). Some literature refers to smaller gang units as "modular forms," and to the larger units as "gang forms."

Gang forms can be made of aluminum (all-aluminum), plywood face and aluminum frame, plywood face and steel frame, and steel. All-aluminum gang forms consist of aluminum sheathing supported by an aluminum frame along with intermediate stiffeners. The aluminum sheathing can be plain or take the shape of a brick pattern for architectural finish. Aluminum sheathing is not popular because of its relatively higher cost and the tendency of concrete to react chemically with aluminum. A common module for all-aluminum gang is 3 × 8 ft (0.91 × 2.44 m) panels.

A more popular and widely used alternative to the all-aluminum gang forms is the aluminum frame with plywood. This system is lighter and less expensive than the all-aluminum gang form. Plywood is attached to the aluminum frame by aluminum rivets.

Another method of attaching plywood to aluminum beams is to use the nailer-type joists in the assembly of the gang form. The plywood is nailed to the nailer type beam by regular nails. A common module for this system is 2 × 8 ft (0.61 × 2.44 m) panels. Figure 6.4 shows a gang form with aluminum frame and plywood face.

The third type of gang forms consists of a plywood face supported by steel walers. Walers are typically made from double channels to allow ties to be inserted between the channels and to reduce the deflection of the gang form. The advantage of this sys-

Figure 6.4   Aluminum frame gang form.

tem over the above mentioned systems is its ability to carry greater loads at longer distances between walers. A common module for this system is the 4 × 8 ft (1.22 × 2.44 m) panel.

The all-steel gang form is made of steel sheathing and steel studs and wales. This system is used to support fresh concrete for high, thick, and multiple lifts. This system has an unlimited number of reuses as long as good storage practices are followed. A common module for this system is the 2 × 8 ft (0.61 × 2.44 m) panel because of its heavy weight.

6.3.2   Gang Forms Assembly

Ganged forms are assembled on the ground, raised into place, and stripped as one unit. Assembly of gang forms starts by placing the walers above lumber blocks on flat and level ground. For faster and more efficient assembly, a gang assembly table can be used instead of assembling the gang on the ground. Walers are then leveled, aligned, and locked in their proper position. The nailer-type beams are then placed on, and perpendicular to, the walers. The nailer-type beams are attached to the walers by clips. Two lumber-end pieces are then placed and attached to the walers. The plywood is then placed and fastened by screws. Tie rod holes can be placed on the ground; however, it is good practice to drill holes and insert tie rods when gang forms are erected to ensure that holes on the two sides of plywood are matched.

6.3.3   Economy and Advantages of Gang Formwork

1.   Productivity of gang forms is higher than traditional forms because they are assembled on the ground and stripped as one unit.
2.   Gang forms produce high-quality smooth concrete with fewer joints. Also, form liners can be attached on the plywood to produce architectural concrete.
3.   Gang forms have higher reuse value than traditional all-wood formwork systems. Also, plywood can be replaced without any need to replace the supporting frame.

6.3.4   Limitations of Gang Formwork

1.   The major limitation of gang formwork is that before moving gang forms vertically or horizontally to the next pouring position, they have to be brought down to the ground for cleaning and oiling. This process substantially increases the cycle time between two lifts.
2.   Gang forms are not suitable for small walls or walls interrupted by pilasters or counterforts.
3.   Because of their large sizes, safety is a major concern when moving ganged forms.

6.4   JUMP FORMS

Jump form systems are used where no floor is available on which to support the wall formwork, or the wall and column proceed ahead of the floor. Jump forms consist of a framed panel attached to two or more strongbacks. They can be one-floor high, supported on inserts set in the lift below, or two sets can also be used, each one-floor high that alternately jump past each other (Figure 6.5).

6.4.1 Jump Form Components

Jump forms consists of two parts: an upper framed panel form with its supporting system and working platform, and a supporting structure that is attached to the concrete wall below the wall being placed. The function of the upper framed panel form is to support the freshly placed concrete. The supporting structure is attached to a stiff concrete wall. Its function is to support the upper framed panel form. Jump form components are shown in Figure 6.6.

Upper Framed Panel Form

The upper part consists of three main elements: (1) framed panel form, (2) supporting brace, and (3) working platforms. The framed panel form consists of a plywood face supported by two or more strongbacks. The frame panel form and the stongbacks are supported by an adjustable pipe brace. The brace is used for plumbing and stripping of the frame panel form. The strongback beams and the pipe brace are rested and connected to a horizontal beam that is anchored to the top of the concrete wall underneath the wall being poured. The strongbacks, brace, and the horizontal beam are forming a truss system that supports the freshly placed concrete. Another function of the horizontal beam is to support the walkway under the lower working platform.

After the concrete gains enough strength to support its own weight, the framed panel form is moved away from the concrete wall to allow the attachment of landing brackets for the next pouring position and to finish concrete patching. The framed panel form is moved away by either tilting or moving horizontally by rollers away from the concrete wall.

There are two working platforms in the upper framed panel form. The upper working platform is used to place and vibrate con-

Figure 6.5 Jump form.

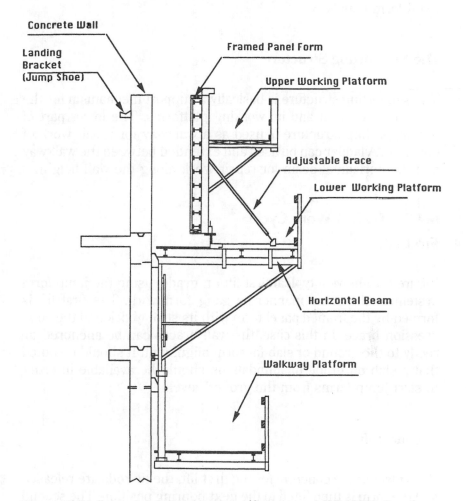

Figure 6.6 Jump form components. (Courtesy of SYMONS Corp.)

crete and to attach the landing bracket (jump shoe). The lower working platform has two functions: (1) to allow construction workers to remove form ties and anchor bolts and (2) to clean and re-oil form panels.

The Supporting Structure

The supporting structure is basically a support mechanism for the framed panel form and its working platforms. The lower part of the supporting structure is used as a walkway for repair work of concrete. A ladder can be used and extended between the walkway and the horizontal beam for repair work along the wall height.

6.4.2   Typical Work Cycle

First Lift

Figure 6.7 shows a typical first lift on grade, using the jump form system in the same manner as gang formwork. The first lift is formed by the framed panel form with its strongback and the compression brace. In this case, the wall braces can be anchored directly to the ground or slab for form alignment. It should be noted that a slab on grade or foundations should be available in order to start jump forms from the ground level.

Second Lift

After placing the concrete for the first lift, the tie rods are released and the form is then lifted to the next pouring position. The second lift begins by attaching the jump shoe to the wall at the first "jump" elevation. The framed panel form is attached to the crane slings and hoisted into position above the jump shoes. The lower supporting structure is then attached without the lower overhanging walkway (Figure 6.8).

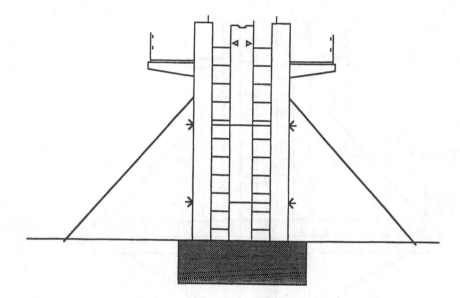

Figure 6.7 Jump form first lift. (Courtesy of Patent Scaffolding Co.)

Third Lift: Walkway Platform Assembly

A finishing walkway platform is added to the jump form. The jump form is now complete for all subsequent lifts. The purpose of the walkway platform assembly is to provide the worker access to jump shoes, wind anchors, and wall patching and finish. It is recommended that when this walkway platform is used, it should be attached after the pour at the first jump position, but prior to raising the form to the third position (Figure 6.9).

Stripping

Stripping begins by removing all form ties and anchor-positioning bolts. The form panel with its strongbacks is then pulled away from the wall by tilting or rolling (Figure 6.10). Tilting is accomplished by releasing the wall brace, while rolling is accomplished by rollers. It should be noted that the compression brace is used for both

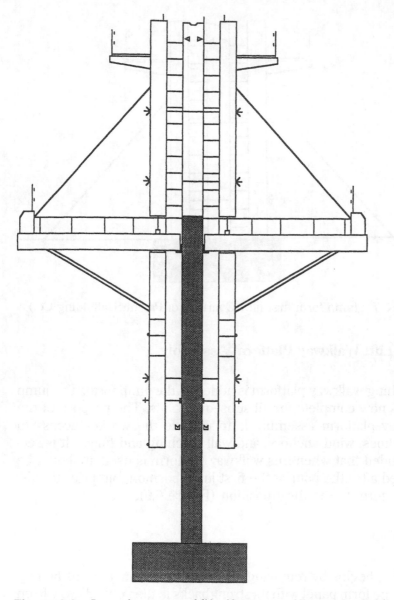

Figure 6.8 Jump form second lift. (Courtesy of Patent Scaffolding Co.)

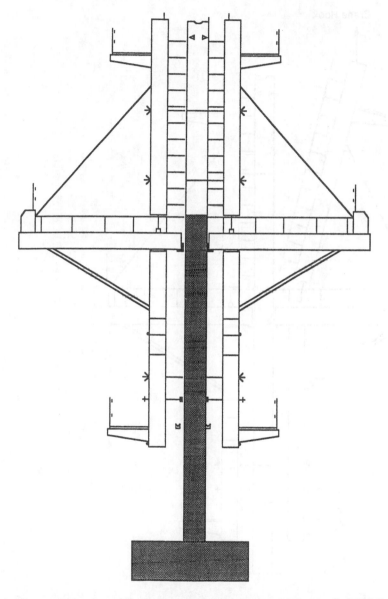
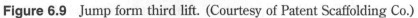

Figure 6.9 Jump form third lift. (Courtesy of Patent Scaffolding Co.)

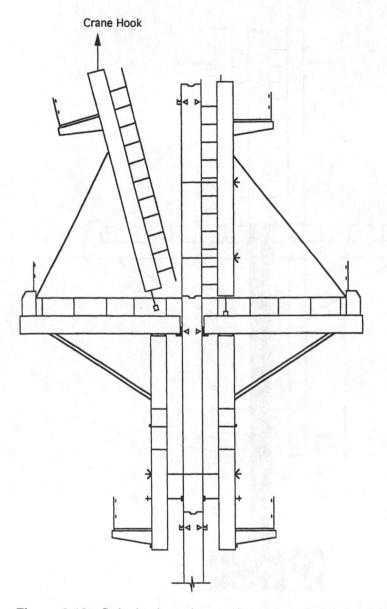

Crane Hook

Figure 6.10   Stripping jump forms. (Courtesy of Patent Scaffolding Co.)

exact adjustment (plumbing) of the form elements as well as for form removal. Stripping allows the inserts to be accessible from the upper level platform. The jump shoes are then positioned and attached for the next lift.

Additional walls or form cleaning, oiling, and repair can be performed from the walkway platform and the lower working platform.

Flying

The entire jump form assembly is then hoisted by crane into position above the newly placed jump shoes (Figure 6.11). The crane lines are attached to the gang form lift brackets at the top of the form panel. The gang is now ready to be set for the next pour.

Resetting

After the crane is released, wind anchors are attached at either the tie location or the jump shoe inserts, and the gang is cleaned and oiled in preparation for the next pour. The form panel is then moved forward until it comes in contact with the top of the previous pour. The gang is plumbed using the wall brace, another gang is positioned on the opposite side of the wall, and ties are installed (Figure 6.12).

It should be noted that there are two different scenarios for forming and pouring concrete walls. First, the slab pour immediately follows each wall, and thereby provides the means of support for the gang on the left side as indicated. Second, there is no floor available to support the formwork, or the walls and columns proceed ahead of the floor. In this case, two sets can be used that jump past each other in an alternate fashion.

6.4.3   Advantages of Jump Form

Significant Reduction in Crane Time

Jump forming can reduce expensive crane time by less than one-half of that required for conventional gang forming used for wall

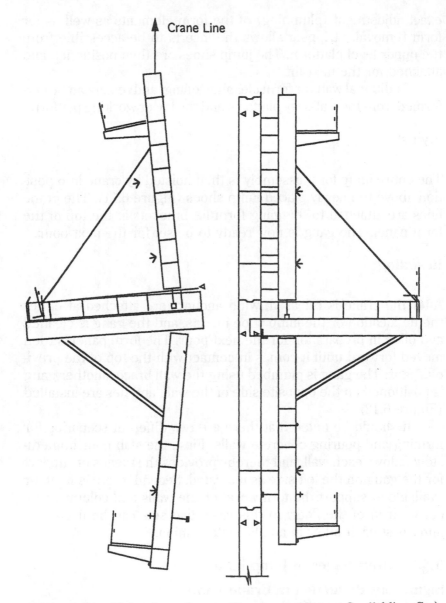

Crane Line

Figure 6.11   Flying jump forms. (Courtesy of Patent Scaffolding Co.)

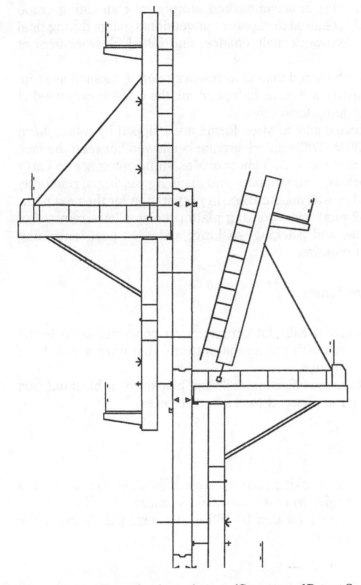

Figure 6.12   Resetting jump forms. (Courtesy of Patent Scaffolding Co.)

construction. This is accomplished simply by eliminating crane time normally required to support conventional gangs during final tie removal, formwork maintenance, and initial tie placement at the next lift.

It should be noted that all formwork operations such as stripping and resetting are crane independent; the crane is only needed for flying the jump form upward.

As indicated above, when forms are stripped by rolling back or tilting, a 30-in. (762-mm) clearance is allowed between the face of the form and the wall. This provides sufficient space to carry out such work as form stripping and cleaning, setting of reinforcement, and other wall maintenance in preparation for the next pour. A 30 in. (762 mm) wide working platform also allows removal of the jump shoe and patching, sacking, and even post tensioning operations if required.

Use of Form Liners

If the specification calls for architectural concrete, jump forms allow form liners with numerous patterns that offer a variety of architectural finishes.

Also, if the specification calls for blockouts, a blockout box is attached to the plywood to create the recess.

Flexibility

Form panels and working platforms are adjustable to achieve any required tilt angle, whether forward or backward. Also, sloped walls can be accommodated by tilting the form panels using the compression brace.

Quality

An aesthetically pleasing concrete finish is always of prime importance in architectural concrete construction such as walls and col-

umns. Jump forms can be complemented with a wide range of form
liners that provide different material types and many different architectural finishes. Depending on the desired finish, these form
liners can be used for producing difficult textures, maximizing reuses, easy stripping, economical single uses, and so on.

Also, because a jump form is a factory built system, it provides
predictable strengths, thereby minimizing the uncertainty that often surrounds equipment fabricated on site.

Durability

When properly anchored, typical jump form applications are designed to withstand wind conditions up to 90 mi (145 km) per hour
and support platform loads up to 50 lb/ft^2 (244.13 kg/m^2). While
the platform is not intended for storage of rebar and other construction materials, this high load capacity does allow the advantage of carrying out reinforcement installation and other general
activities directly from the platform.

Safety

The wide (5 to 6 ft) (1.52 to 1.83 m) guarded working platforms
provide a very secure work area for construction crews. Also, it is
not necessary for any crew member to be on the form during crane
handling to assist in the lift procedure.

Productivity

The jump form system is a very productive one that allows contractors to complete a floor cycle every 2 to 4 days, depending on the
size of the floor and the height of the wall. Also, because jump
forms are braced from the outside, no or minimum inside bracing
is needed, thus eliminating interference with interior shoring.

6.4.4   Limitations of Jump Form
Accessibility

The site must be fairly accessible, since the forms can be up to
16 ft (4.88 m) high and 44 ft (13.41 m) long.

Openings/Inserts

Jumps forms are best suited to building designs in which the openings are regularly occurring from floor to floor. The existence of
openings, blockouts, and inserts slows the jump form operation.

Clearance

Free space is required between the forms and an adjacent building
in order to advance from one floor to the next.

7
Wall Form Design

This chapter presents a design method for all-wood concrete wall forms. This procedure was formulated to provide for a safe wall form design for all components. The design methodology is based on loads recommended by ACI-347-1994 and stresses values recommended by NDS-1991 and APA 1997.

7.1 WALL FORM COMPONENTS

A wall form is usually made up of sheathing, studs, wales, ties, and bracing as shown in Figure 6.1. The fresh concrete places a lateral pressure on the sheathing, which is supported by studs. Studs behave structurally as a continuous beam with many spans supported on wales. Wales, in turn, are assumed to act as a continuous beam that rests on ties. Ties finally transmit concrete lateral pressure to the ground.

7.2 DESIGN LOADS

The pressures exerted on wall forms during construction need to be carefully evaluated in the design of a formwork system. Loads imposed by fluid concrete in walls and columns are different from gravity loads produced on slab forms. Fresh concrete exhibits temporary fluid properties until the concrete stiffens sufficiently to support itself.

7.2.1   Lateral Pressure of Concrete Forms for Wall

Formwork should be designed to resist the lateral pressure loads exerted by the newly placed concrete in the forms. If concrete is placed rapidly in wall or column forms, the pressure can be equivalent to the full liquid head pressure. This requires that rate of placement exceed the initial set time of the concrete mix. Excessive deep vibration can liquefy the initial set of concrete within the effective coverage of the vibrations. Retarder additives or cool weather can also delay the initial set and result in higher than anticipated lateral pressure. The formula for wall pressure established by the American Concrete Institute (ACI-347) considers the mix temperature and the rate of placement of concrete. The rate of placement is expressed in terms of feet per hour of concrete rise in the forms. Table 7.1 shows pressure values for concrete walls of different temperature and rate of filling.

Table 7.1   Pressure Values for Concrete Walls: Relation among the Rate of Filling Wall Forms, Maximum Pressure, and Temperature (ACI)

Rate of filling forms, ft/h	Maximum concrete pressure, lb/ft^2						
	Temperature, °F						
	40	50	60	70	80	90	100
1	375	330	300	279	262	250	240
2	600	510	450	409	375	350	330
3	825	690	600	536	487	450	420
4	1050	870	750	664	600	550	510
5	1275	1050	900	793	712	650	600
6	1500	1230	1050	921	825	750	690
7	1725	1410	1200	1050	933	850	780
8	1793	1466	1246	1090	972	877	808
9	1865	1522	1293	1130	1007	912	836
10	1935	1578	1340	1170	1042	943	864
15	2185†	1858	1573	1370	1217	1099	1004
20	2635†	2138†	1806	1570	1392	1254	1144

† These values are limited to 2000 lb/ft^2.

1. For columns and walls with rate of placement less than 7 ft/h (2.1 m/h);

    $$p = 150 + \frac{9000R}{T}$$

    with a maximum of 3000 psf (1.47 kgf/cm²) for columns, 2000 psf (0.98 kgf/cm²) for walls, a minimum of 600 psf (0.29 kgf/cm²), but no greater than 150h (0.24h_{st}).

    where

    p = lateral pressure (lb/ft²)
    R = rate of placement, ft/h
    T = temperature of concrete in the form °F
    h = height of the form, or the distance between construction joints, ft

    or

    $$p_M = 0.073 + \frac{8.0R_{st}}{T_C + 17.8} \qquad \text{(metric equivalents)}$$

    where,

    P_M = lateral pressure, kgf/cm²
    R_{st} = rate of placement, m/h
    T_c = temperature of concrete in the forms, °C
    h_{st} = height of fresh concrete above point considered, m

2. For walls with rate of placement of 7 to 10 ft/h (2 to 3 m/h):

    $$p = 150 + \frac{43,400}{T} + \frac{2800R}{T}$$

or

$$p_M = 0.073 + \frac{11.78}{T_c + 17.8}$$

$$+ \frac{2.49R_{st}}{T_c + 17.8} \quad \text{(metric equivalents)}$$

with a maximum of 2000 psf [0.98 kgf/cm^2], a minimum of 600 psf [0.29 kgf/cm^2], but no greater than 150h ($0.24h_{st}$).

3. For rate of placement > 10 ft/h:

$$p = 150h$$

or

$$p_M = 0.24h_{st} \quad \text{(metric equivalents)}$$

The above three formulas can only be applied if concrete satisfies the following conditions:

-   Weighs 150 pcf (2403 kg/m^3)
-   Contains no admixtures
-   Has a slump of 4 in. (100 mm) or less.
-   Uses normal internal vibrator to a depth of 4 ft (1.22 m) or less

If concrete is pumped from the base of the form, the form should be designed to resist the lateral hydrostatic pressure of fresh concrete plus minimum allowance of 25 percent to account for pump surge. Caution must be taken when using external vibration or concrete made with shrinkage-compensating or expansive cements as pressure higher than the hydrostatic pressure is expected to occur. It is a good practice to reduce the allowable stresses to half its original value when using external vibrators.

7.2.2 Horizontal Loads

Braces should be designed to resist all foreseeable horizontal loads, such as seismic forces, wind, cable tension, inclined supports, dumping of concrete, etc.

Wall form bracing must be designed to meet the minimum wind load requirements of ANSI A58.1 or the local design building code, whichever is more stringent. For exposed wall, the minimum wind design load should not be less than 15 psf. Bracing for wall forms should be designed for a horizontal load of at least 100 lb per lineal foot of the wall, applied at the top.

7.3 METHOD OF ANALYSIS

Step 1:   The procedure for applying equations of Tables 3.15 and 3.16 to the design of a sheathing is to consider a strip of 1 ft depth (consider the lower 1 ft of sheathing where concrete lateral pressure is maximum). Determine the maximum allowable span based on the allowable values of bending stress, shear stress, and deflection. The lowest value will determine the maximum spacing of studs.

Step 2:   Based on the selected stud spacing, the stud itself is analyzed to determine its maximum allowable spacing. The studs are subject to uniform pressure resulting from the fresh concrete. This pressure is resisted first by the sheathing which in turn transfer the loads to studs. The selected stud span will be the spacing of the wales.

Step 3:   Based on the selected stud spacing, the maximum wale spacing (distance between horizontal supports or ties) can be determined using the same procedure. For simplicity and economy of design, this maximum span value is usually rounded down to the next lower integer or modular value when selecting the spacing.

7.4 STRESSES CALCULATIONS

After appropriate design loads are calculated, the sheathing, studs, and wales are analyzed in turn, considering each member to be a uniformly loaded beam supported in one of the three conditions (single span, two spans, or three or more spans) to determine the stresses developed in each member. Vertical supports and lateral bracing must be checked for compression and tension stresses. Except for sheathing, bearing stresses must be checked at supports to ensure safety against buckling. Using the methods of engineering mechanics, the maximum values of bending moment, shear, and deflection developed in a uniformly loaded beam of uniform cross section is shown in Tables 3.15 to 3.17.

The maximum fiber stresses in bending, shear, and compression resulting from a specified load may be determined from the following equations:

Bending: $f_b = \dfrac{M}{S}$

Sheer: $f_v = \dfrac{1.5V}{A}$ for rectangular wood members

$\quad\quad\quad = \dfrac{V}{lb/Q}$

Compression: f_c or $f_{c\perp} = \dfrac{P}{A}$

Tension: $f_t = \dfrac{P}{A}$

where $f_b, f_v, f_{c\perp}, f_c,$ and f_t are as defined before in the NDS tables, and

$$A = \text{section area, in.}^2$$
$$M = \text{maximum moment, in.-lb}$$
$$P = \text{concentrated load, lb}$$
$$S = \text{section modules, in.}^3$$
$$U = \text{maximum shearing force, lb}$$
$$lb/Q = \text{rolling shear constant, in.}^2/\text{ft}$$

7.5   DETERMINATION OF MAXIMUM ALLOWABLE SPAN

Maximum span corresponding to bending, shear, and deflection can be directly obtained using equations given in Table 3.16. As previously mentioned, the maximum allowable design value for the span will be the smallest one rounded it to the next lower integer or modular value.

7.6   DESIGN OF LATERAL BRACING

For wall forms, lateral bracing is usually provided by inclined rigid braces. Bracing is usually required to resist wind loads and other horizontal loads. Since wind load may be applied in either direction, braces must be arranged on both sides of the forms. When rigid braces are used, they may be placed on one side of the form if designed to resist both tension and compression. Figure 7.1 gives a visual example of form bracing.

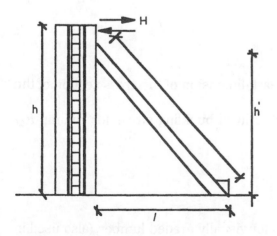

Figure 7.1   Bracing of formwork.

Design Load

Design load for bracing can be calculated using the following equations:

$$P' = \frac{Hhl}{h'l'} \qquad l = (h'^2 + l'^2)^{1/2}$$

where

P' = strut load per foot of the form, lb/ft
H = lateral load at the top of the form, lb/ft
h = height of the form, ft
h' = height of the top of strut, ft
l = length of strut, ft
l' = horizontal distance from bottom of strut to form (ft)

Design Procedure

1. Start design by selecting a certain strut size such that it satisfies

$$\frac{l}{d} \leq 50$$

where d is the least dimension of the cross section of the selected strut.

2. Calculate Euler's critical buckling stress for column F_{CE} as follows:

$$F_{cE} = \frac{K_{cE}E'}{(l_e/d)^2}$$

where $K_{cE} = 0.3$ for visually graded lumber (also used in form design)

3. Calculate the limiting compressive stress in column at zero slenderness ratio F_c^* from the equation:

$$F_c^* = F_c(C_D)(C_M)(C_t)(C_F)$$

where C_D, C_M, C_t, C_F are defined tables (see Tables 3.4a,b, 3.7, and 3.8)

4. Calculate the column stability factor C_p from the formula:

$$C_p = \frac{1 + F_{cE}/F_{cE}^*}{2 \times 0.8} - \sqrt{\left(\frac{1 + F_{cE}/F_{cE}^*}{2 \times 0.8}\right)^2 - \frac{F_{cE}/F_{cE}^*}{0.8}}$$

5. The allowable compressive stress F_c' in the strut is given by

$$F_c' = F_c^*(C_P)$$

6. If $F_c' < F_{cE}$, this means that the selected cross section is not enough to resist buckling. So increase the size of the cross section and go iteratively through steps 1 to 6 until you get $F_{cE} < F_c'$.

7. The maximum load that can be carried by the strut is the product of F_c' and the actual (not the nominal) cross-sectional area of the selected strut.

8. The maximum spacing of struts in feet that can be carried by one strut is obtained by dividing the maximum load by strut load per foot.

It should be noted that the strut usually carries compression or tension force depending on the direction of the horizontal load applied to the form. Those two forces are equal in magnitude but differ in their sign. Designing struts as compression members usually ensures that they are safe also in tension because we are considering an additional precaution against buckling associated with compression.

EXAMPLE 1

Design formwork for a 15-ft-high concrete wall, which will be placed at a rate of 4 ft/h, internally vibrated. Anticipated temperature of the concrete at placing is 68°F. Sheathing will be 1 in. thick (nominal) lumber and 3000-lb ties are to be used. Framing lumber is specified to be of construction grade Douglas Fir No. 2.

SOLUTION:

Design Load

Lateral pressure:

$$P = 150 + \frac{9000R}{T}$$

$$P = 150 + \frac{9000 \times 4}{68}$$

$$= 679.41 \text{ psf} \cong 680.0 \text{ psf}$$

Assume construction joint every 5 ft:

$$150 \times h = 150 \times 5 = 750 \text{ psf}$$

$$680 < 3000 \quad \text{and} \quad 680 < 150 \times h \quad \text{(OK)}$$

The design value for lateral pressure is 680 psf.

Design Criteria

One needs to find the maximum practical span that the design element can withstand.

Stud Spacing

Consider 12-in. strip.
Load/ft' = 680 lb/ft^2

From design tables we can get:

- $F_b = 875$ psi
- Flat use factor $C_{fu} = 1.2$
- Size factor $C_f = 1.2$
- $F_v = 95$ psi (here we have no split)
- Temperature factor $= C_t = 1.0$
- Load duration factor $= C_D = 1.25$

(Load duration $= 7$ days for most formwork unless otherwise stated.)

Bending

Allowable stress $= F_b' = F_b(C_{fu})\,(C_t)\,(C_f)\,(C_D)$
$F_b' = 875 \times 1.2 \times 1.0 \times 0.9 \times 1.25 = 1181.25$ psf
   $l =$ allowable span

$$= 10.95 \left(\frac{F_b'}{w}S\right)^{1/2} = 10.95 \left(\frac{1181.25 \times 1.055}{680}\right)^{1/2}$$

$$= 14.824 \text{ in.}$$

Shear

Allowable stress $= F_v' = F_v(C_H)\,(C_t)\,(C_D)$
$F_v' = 95 \times 2.0 \times 1.0 \times 1.25 = 237.5$ psi
$l = 13.3 \left(\frac{F_v'A}{w}\right) + 2 \times d$

From tables we can get:

- $d = 0.75$ in.
- $A = 8.438$ in.^{2}

$$l = \frac{13.3 \times 237.5 \times 8.438}{680} + 2 \times 0.75 = 40.7 \text{ in.}$$

Deflection

$$l = 1.69 \left(\frac{EI}{w}\right)^{1/3}$$

From tables we can get:

- $I = 0.396$ in.^{4}
- $E = 1,600,000$ psi

$$E' = E\ (C_t) = 1.6 \times 10^6 \times 1.0 = 1.6 \times 10^6 \text{ psi}$$

$$l = 1.69 \left(\frac{1.6 \times 10^6 \times 0.396}{680}\right)^{1/3} = 16.507 \text{ in.}$$

Hence sheathing will be supported by studs, with a spacing of 12 in. (1 ft).

Wale Spacing:

Load/ft′ = w × (stud spacing) × 1 ft' of wale span
$$= 680.0 \times 1.0 \times 1.0 = 680.0 \ lb/ft$$

Try 1 (2 × 4) Douglas Fir.
From tables we can get:

- $I = 5.358$ in.^{4}
- $S = 3.063$ in.^{3}
- $d = 3.5$ in.
- $A = 5.25$ in.^{2}

Bending

Size factor $C_f = 1.5$
F'_b = allowable stress = $F_{b(\text{studs})}\ (C_t)\ (C_D)\ (C_f)$

$$F'_b = 675 \times 1.0 \times 1.25 \times 1.5 = 1266.0 \text{ psi}$$

$$l = 10.95 \left(\frac{1266.0 \times 3.063}{680} \right)^{1/2} = 26.150 \text{ in.}$$

Shear

$$F'_v = F_v(C_H)(C_t)(C_D)$$
$$F'_v = 95.0 \times 2.0 \times 1.0 \times 1.25 = 237.5 \text{ psi}$$

$$l = 13.3 \frac{F'_v A}{w} + 2d$$

$$l = 13.3 \times \frac{237.5 \times 5.25}{680} + 2 \times 3.5 = 31.387 \text{ in.}$$

Deflection

As before, $E' = 1.6 \times 10^6$ psi.

$$l = 1.69 \left(\frac{1.6 \times 10^6 \times 5.359}{680} \right)^{1/3} = 39.336 \text{ in.}$$

Bending governs; span = 26.150 in. (take 2 ft). Final stud spacing is 2 ft.

Double Wales

Try 2 (2 × 4) Douglas Fir (properties are same as above).

$$\text{Load}/f' = 680 \times \text{(stud spacing)}$$
$$\times 1 \, ft' \text{ of wales} = 1360 \text{ lb/ft}$$

Bending

$$F_b' = \text{allowable stress} = F_{b\,(\text{studs})}(C_t)(C_D)(C_f)$$

$$F_b' = 875.0 \times 1.0 \times 1.25 \times 1.5 = 1640.0 \text{ psi}$$

$$l = 10.95 \left(\frac{1640.0 \times 2 \times 3.063}{1360}\right)^{1/2} = 29.762 \text{ in.}$$

Shear

$$F_v' = F_v(C_H)(C_t)(C_D)$$

$$F_v' = 95.0 \times 2.0 \times 1.0 \times 1.25 = 237.5 \text{ psi}$$

$$l = 13.3 \left(\frac{F_v' \cdot A}{w}\right) + 2 \times d$$

$$l = 13.3 \left(\frac{237.5 \times 5.25}{680}\right) + 2 \times 3.5 = 31.387 \text{ in.}$$

Deflection

As before, $E' = 1.6 \times 10^6$ psi.

$$l = 1.69 \left(\frac{1.6 \times 10^6 \times 5.359}{680}\right)^{1/3} = 39.336 \text{ in.}$$

Bending governs; span = 29.762 in. (take 2.5 ft). Practically, we sometimes round this spacing up to 2.5 ft as 29.762 is very close to 30.

The wales will be supported by bracing, with a spacing of 2 ft.

Tie Spacing

- 3000-lb ties will be used.
- Force/tie = (lateral concrete pressure) × (wale spacing) × (tie spacing), which should not be greater than 3000 lb.
- $3000 = 680 \times 2 \times$ tie spacing $\rightarrow$ tie spacing ≤ 2.206 ft.

Take tie spacing = 2 ft.

Bearing Stress

1. Studs on wales:
   Bearing area $= b_{wale} \times b_{stud} \times 2.0$
   (we multiply by 2 because of the double wale).
   Load transmitted by bearing = pressure intensity × stud spacing × wale spacing $= 680.0 \times 1.0 \times 2.0 = 1360.0$ lb
   From tables we can get:

   - Bearing area factor $= C_B = 1.25$
   - Temperature factor $= C_t = 1.0$
   - $F_{C\perp} = 625.0$ psi

   $F'_{C\perp} = F_{C\perp} (C_B) (C_t)$
   $F'_{C\perp} = 625 \times 1.25 \times 1.0 = 781.25$ psi

   The calculated value of bearing stress is:

   $$F_{C\perp \text{(calculated)}} = \frac{1360.0}{(1.5)^2 \times 2} = 302.22 < 781.25 \quad \text{(safe)}$$

2. Tie plate:

Tie load = (pressure intensity) × (tie spacing)
$$\times \text{ (wale spacing)} = 680.0 \times 2 \times 2 = 2720.0 \text{ lb}$$

Allowable stress in bearing is 781.25 psi.

Area of bearing shown in the figure =
(width of wale) × (required width of tie plate)
× 2 (since we have 2 wales)

B, the required width of the tie plate, can be obtained as follows:

$$2 \times 1.5 \times B = \frac{2720.0}{781.25}$$

$B \geq 1.161$ in.; take $B = 1.5$ in.

Strut Load

Try 4 × 4 in.

$H = 100$ lb/ft
$h = 15$ ft
$h' = 10$ ft
$l' = 10$ ft
$l = 14.14$ ft

$$P' = \frac{H \times h \times l}{h' \times l'} = \frac{100 \times 150 \times 10\sqrt{2}}{10 \times 10} = 2121.30 \text{ lb}$$

$$\frac{l}{d} = \frac{14.14 \times 12}{3.45} = 48.48 < 50.0 \qquad \text{(OK)}$$

$$F'_{cE} \text{ in strut} = \frac{0.3E'}{(l/d)^2} = \frac{0.3 \times 1.6 \times 10^6}{(48.48)^2}$$

$$= 204.228 \text{ psi}$$

$F_c^* = F_c \times$ (all factors except C_p)

$F_c^* = F_c(C_D)(C_M)(C_t)(C_F)$

$F_c^* = 1300.0 \times 1.25 \times 1.0 \times 1.0 \times 1.5 = 2437.5 \text{ psi}$

C_p = column stability factor

$$= \frac{1 + \dfrac{204.228}{2437.5}}{2*0.8} - \sqrt{\left(\dfrac{1 + \dfrac{204.228}{2437.5}}{2*0.8}\right)^2 - \dfrac{\dfrac{204.228}{2437.5}}{0.8}} = 0.0823$$

$F'_c = F_c^*(C_P) = 2437.5 \times 0.0823 = 200.629 \text{ psi}$ \qquad (unsafe)

Try 5×5 in. Following the same procedure, we get

$$\frac{l}{d} = 37.7067 \qquad F_{cE} = 337.6014 \text{ psi}$$

$C_P = 0.03784 \Rightarrow F'_c = 922.2754 \text{ psi}$

Allowable P is $922.2754 \times (4.5)^2 = 18676.07$ lb. and spacing between struts $= 18676.07/2121.30 = 8.8041$ ft. Take spacing $= 8$ ft.

EXAMPLE 2

Rework Example 1 with sheathing that has the following characteristics:

Plywood is Type: APA B-B PLYFORM CLASS 1 with species group of face ply = 1.

- Dry condition.
- Thickness: $7/8$ in.

SOLUTION:

From Tables 3.11 and 3.12 of geometric properties, we get:

- KS = 0.515 in.
- I = 0.278 in.^{4}
- lb/Q = 8.05 in.^{2}

From Table 3.14 of mechanical properties, we can obtain the following design values:

- F_b = 1650.0 psi.
- F_v = 190.0 psi
- E = 1.8 $\times$ 10^6 psi

Concrete pressure is 680.0 psf. [see Example 1]

STUD SPACING:

Bending

$$w_b = 680 = 120 \times F_b \times \frac{KS}{l_1^2}$$

$$680 = \frac{120 \times 1650.0 \times 0.515}{l_1^2}$$

$$l_1 = 12.246 \text{ in.}$$

Shear

$$w_s = 20 \times F_v \times \frac{lb/Q}{l_2}$$

$$680.0 = \frac{20.0 \times 190.0 \times 8.05}{l_2}$$

$$l_2 = 44.985 \text{ in.}$$

Deflection

Due to bending only

$$\Delta \leq \frac{l}{360}$$

$$\Delta = \frac{wl_3^4}{1743EI}$$

$$\frac{l_3}{360} = \frac{680 \times l_3^4}{1743 \times 1.8 \times 10^6 \times 0.278}$$

$$l_3 = 15.273 \text{ in.}$$

Bending governs. Stud spacing = 12.246 in. rounded down to 12 in. = 1 ft. The rest of the design is as before.

8

Vertical Formwork Systems: Crane-Independent Systems

As mentioned in Chapter 5, vertical formwork can be classified into two main categories, namely, crane dependent systems and crane-independent systems. Slipform and self-raising formwork are classified as crane-independent systems in which formwork panels are moved vertically by other vertical transportation mechanisms.

8.1 SLIPFORMS

8.1.1 History

The history of slipforming can be traced back to the 1920s and before. In the late 1920s, a number of concrete structures were cast using a system of formwork that was moving during the placing of concrete. At that time, forms were being raised by hand-screw jacks and job-built wooden yokes. Laborers would pull ropes that would raise the forms. Early application of slipform was limited to storage bins and silos with a constant thickness all over the wall height.

Since the late 1950s, slipform construction has come a long way; locomotion is accomplished by jacks climbing on smooth steel rods or pipes anchored at the base of the structure and the system has been employed successfully and economically in situations which have required discontinuity of section. As a result, the list of recent application expanded to include towers cores, bridge piers, power plant cooling, chimney shafts, pylons, and the legs of oil rig platforms.

8.1.2 Construction Practices

Most contractors using slipforming do not own their own slipforming equipment. A general contractor who finds it feasible to use slipforming has three options:

1. To design, build, and operate the slipform complex. This requires the assistance of expert consultants who will help at every step of the project.
2. To subcontract the work to a subcontractor who specializes in slipform.
3. To buy or rent the forming system completely designed and fabricated, ready to be erected at the site to be operated by the contractor's own forces. The cost of renting slipform equipment is based on the linear feet (or meters) of a slide, the capacity and the quality of the equipment. Jacks and jacking rods are used on a rental basis but forms aren't normally available for rental.

8.1.3 The Slipforming Operation

Slipform construction is an extrusion process in which the form, 3.5 to 6 ft (1.07 to 1.83 m) high, is the die and is constantly being raised. Fresh concrete is placed or pumped into the forms. After two to three hours, the concrete reaches the initial set and loses its plasticity and starts supporting the newly fresh concrete above. The rate of movement of the forms is controlled and matches the initial setting of concrete so that the forms leave the concrete after it is strong enough to retain its shape while supporting its own weight. The forms move upward by mean of jacks climbing on smooth steel rods embedded in the hardened concrete and anchored at the concrete foundation base. These jacks may be hydraulic, electric, or pneumatic and operate at speeds up to 24 in./h (609.6 mm/h).

Figure 8.1 shows a slipforming system supported by the jacking rods. Yokes are frames that are used to support the lateral loads and transfer the vertical loads to the jack rods. Workers con-

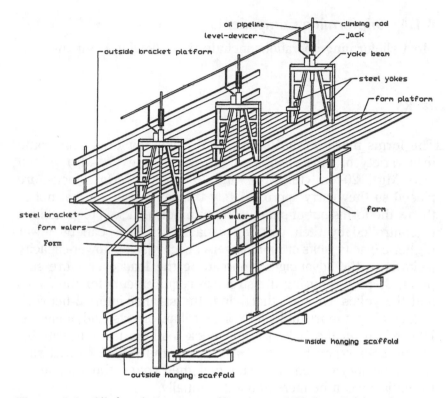

oil pipeline

level-devicer

outside bracket platform

climbing rod

jack

yoke beam

steel yokes

form platform

steel bracket

form walers

Form

form walers

form

inside hanging scaffold

outside hanging scaffold

Figure 8.1   Slipforming system. (Courtesy of Gleibau Salzburg)

tinually vibrate the concrete so as to prevent any honeycombing. An upper working platform is attached to the inner form and slides up with it to provide a place from which workers can place concrete and fabricate steel reinforcement. A lower working platform is suspended from the outer form to allow workers to apply a curing compound and to repair any honeycombing that may occur.

Setting up slipforming starts by placing the jack rods in the foundation. A grid of steel girders is constructed and supported by these jack rods. From this grid of steel girders, the system of sheathing and decks are placed and supported. This process of setting up the slipform takes 3 to 5 weeks.

8.1.4   Slipforming Components

Most slipforming operations include the following components.

Jacks

The forms move upward by mean of jacks. Slipform jacks come in a variety of capacities, 3-ton (2.7-Mg), 6-ton (5.4-Mg), 15-ton (13.6-Mg), 20-ton (18.1-Mg), and 22-ton (20.0-Mg). Jacks are placed so they carry approximately the same load, so that not to throw the forms out of plumb. The number of jacks being utilized is controlled by their carrying capacities. There can be fewer higher-capacity jacks or more lower-capacity jacks. Lower-capacity jacks have the advantage of spreading the load over more supports, thereby reducing the strength requirements for the forms and the yokes. On the other hand, the smaller interval between jacks makes it more difficult to place rebar, inserts, and openings. Depending on the jack size and deck loading, the intervals between jacks range between 4 ft (1.2 m) and 9 ft (2.7 m). Using high-capacity jacks and specially designed yokes, the interval between jacks can be increased substantially.

Jacks come with an individually incorporated self-leveling device. In most cases, keeping the jacks level within ½ in. (12.7 mm) is satisfactory. Jacks are pneumatic, electric, or hydraulic. Hydraulic jacks are relatively light and compact, easy to install, and quite reliable under almost all conditions. Pneumatic jacks operate very much like hydraulic jacks except that they use air instead of oil. They are considerably larger in diameter and have a tendency to malfunction in cold weather. Electric jacks also have the advantage of being automatically self-leveling, as they are activated by a water-level system running through all the jacks with the water level controlled from a control point. Each jack has its own electric motor which operates an electric arm attached to the jaw system. Electric jacks are larger than the hydraulic jacks and require electric and water connections as well as individual oil reservoirs.

Jacking Rods

The jacking rods used in slipforming operations are usually ¾ in. (19.05 mm) pipe, 1¼ in. (31.75 mm) solid rods, or 2½ in. (63.50 mm) pipe, depending on the design capacity. These rods are commonly 10 to 20 ft (3.05 to 6.1 m) with their ends drilled to receive a joining screw dowel connecting one jack rod or tube to the next. Jacking rods can be left in the forms and used as reinforcing. Rods are placed inside the forms and are prevented from buckling by the concrete that has already hardened. Since the jacking rods depend on the concrete for stability, it is necessary to support the rods against buckling when they are out of the concrete, such as when the rods go through a door or window opening. As a result, and whenever possible, jacking rods should be placed to miss any repeated wall openings.

Sheathing

Slipforms consist of inner and outer sheathing (form), 3.5 to 6 ft (1.07 to 1.83 m) high, using 1 in. (25.4 mm) thick lumber. Forms may be fabricated from wood or steel. The sheathing is not fixed to the floor; instead it is suspended either from several lifting devices supported on metal rods, or from other members resting on the foundation or on hardened concrete by means of wooden or metal yokes (frames). Once the form has been filled with fresh concrete and hardening has started, the form is gradually raised by the lifting devices on which it is suspended.

Yokes

Yokes are normally made of steel or wood. The function of yokes is to transfer the entire loads of decks and supporting scaffolding into the jacks and the jack rods.

Form Platform (Working Deck)

A working deck (form platform) is attached to the form and slides up with it to provide a place from which workers can place concrete and fabricate steel reinforcement. Other functions of the working deck include:

1. Placement and vibration of concrete
2. Placement of horizontal and vertical rebar
3. Placement of keyway and dowel anchors for slab (if available) to core connection.

Finish Scaffold

A lower hanging scaffold is suspended from the outer form to allow workers to apply a curing compound and to repair any honeycombing that may occur. The finish scaffold can be also used for stripping of the forms for openings, keyways, and dowels.

8.1.5   Methods and Techniques in Slipform Operation

Method of Placement

Concrete is commonly hoisted by cranes, deck-mounted hoists, or pumping. Deck-receiving hoppers take the concrete in where it is distributed to the various locations in the forms by means of hand buggies. Traffic patterns must be worked out as to allow for easy access of concrete and for the storage of rebar, embedments, and openings. Concrete is usually placed in 2 to 12 in. (50.8 to 304.8 mm) layers, with the 12 in. (304.8 mm) layers the most desirable. The slow rate of placement 2 or 3 in. (50.8 or 76.2 mm) layers is used when there is a delay in the process and the concrete isn't allowed to have any cold joints (silos, bins, and cooling towers).

Vibration

Vibration of concrete is very important to the quality of the concrete. One inch (25.4 mm) diameter vibrators are normally used unless the wall being poured is thick, in which case large vibrators up to 3 in. (76.2 mm) in diameter can be used. The vibration of concrete should penetrate just the top layer. The deep vibration of concrete could cause a fall-out of concrete from under the form. Care should be taken to vibrate every section of concrete. Honeycombs in concrete are a result of poor vibration effort.

Temperature and Concrete Mix

Ambient temperature, speed of slipforming raise, and the conditions under which the concrete is to be placed are among several factors that should be considered when selecting the type and ratio of cement in the concrete mix. Type I normal Portland cement is recommended in almost all instances. Type II modified cement is used for subgrade work and mass concrete pours. Type III high-early-strength cement should only be used in extremely cold temperatures. The design strength for slipform concrete mix should be between 3000 and 4500 psi (2.11 to 3.16 kg/mm^2). The cement content should be between 6 and 7½ sacks per cubic yard (7.8 and 9.8 sacks per cubic meter) of concrete in the summer and between 7 and 8 sacks per cubic yard (9.2 and 10.5 sacks per cubic meter) in winter construction. Below these ratios, harshness, honeycomb, and cold joints will be encouraged because of the stiffness of the mix. A high cement content may generate excessive heat, making the concrete hard to place, entrapping air, and increasing temperature stress in the green concrete. For good placement of concrete it is recommended that the slump of the mix should be about 4 in. (101.6 mm). Temperature is probably the most important factor in good concrete placement and least emphasized.

Other factors that contribute to the quality of concrete include good aggregate size gradation and the use of retarders or accelerators to control the rate of concrete setting.

Speed of Operation

The average speed at which the slipform will be operated must be decided based on the concrete mix, the forms themselves, the storage capacities, and the equipment for raising workers and materials. This consideration will often include decisions as to the use of concrete additives, heating, use of ice, form insulation, enclosures, and so on.

Concrete is normally pumped to the forms at a rate of 18 to 20 cyd/h (16.4 to 18.3 m/h). The forms are normally raised at a rate of 9 to 12 in./h (228 to 305 mm/h) and can reach a rate of 20 in./h (508 mm/h). At this rate they are pouring about a floor a day.

Admixtures

Admixtures can be added to the concrete to speed up or slow down the hydration process of concrete. These should only be used when the specifications call for them or by the request or approval of the architect or field representative. Precision should be exercised in the addition of admixtures because if not carefully controlled, the side effects become more significant as the dosage increases. Admixtures should not be used in lieu of temperature control. Heating or cooling the materials is effective and involves none of the dangers associated with chemical control. Substituting ice for water in the mix is the best method of cooling concrete and is highly recommended. Heating the concrete is usually done by running steam pipes near the forms or by placing electrical or propane heaters near the forms.

Using fly ash in the concrete mix is not recommended because of its tendency to bond with the forms.

Curing/Finishing

Curing and finishing is ideal after the concrete leaves the bottom of the forms. Normally, a float and brush finish is applied, but other means of applying finishes are fast coming into use, a spray finish mainly. Curing the concrete used to be done by spraying the concrete mechanically with water, but this process leaves gouges, misses areas, and creates messy water conditions at the base of the structure. The use of curing membranes or worker-applied spraying compounds is now a more popular form of curing the concrete.

Rebar

The placement of rebar in slipform construction is a difficult task and must be planned very carefully. The need for careful planning of placement of rebar is because the beams of the jacking yokes are moving upward and the horizontal reinforcing steel is stationary, the horizontal rebar must be threaded through the yokes, and the time between concrete pours is short. As a result, the reinforcing should be on the job well in advance to placement, for a shortage of rebar and a shutdown of the slipform could be very costly.

Forming Openings, Projections, and Recesses

Openings, doorways, and ductwork are formed by taking out entire sections of sheathing. Also, keyways, anchor plates, and threaded inserts have to be placed into the forms while the concrete is being poured. The vertical placement of these items is measured by the use of marker rods. These rods are separate from jacking rods and rebar but are placed in the concrete at the beginning of the slipforming. They are used to mark the vertical progress of the job and to indicate when to place any type of inserts or blockouts. The horizontal placement of these blockouts and inserts are indicated on the inside sheathing of the forms themselves. Different colors of tape are used to indicate the center of different types of blockouts and inserts. These blockouts and inserts are tied to the rebar with No. 9 wire to prevent their movement while in the forms. When blockouts are to be placed where

jacking rods are, the rods must be adequately braced against buckling. Repeated blockouts are a major design factor in the placement of the jacking rods.

Reducing Wall Thickness

In tall structures, it is necessary to reduce the wall thickness peri-
odically to reduce the weight of the structure and to save in the
volume of concrete. Whenever a reduction in wall thickness is required, a board is placed on the inside of the forms thereby reducing the thickness of the wall. When the reduction is on only one
side of the wall, the jacks will not be in the center of the forms
and the pull on the yokes must be balanced.

Leveling

One of the major problems with the slipform construction is keeping the forms plumb and level. The most basic and often over-
looked means of preventing misalignment is adequate bracing. All
the forms, decks, and jacking equipment should be securely fastened to the jacking grid. Also the placement of jacks is important
to ensure that the forms are raised in a uniform manner.

A water system made of a centralized reservoir and open-
ended tubes at various yokes in the system and large plumb bobs
suspended by piano wire are commonly used to check if the building is plumb and level. More recently, vertical lasers placed at two
or three corners of the building are being used more often. Also,
transits or theodolites are used to check for alignment. If and when
the building becomes out of plumb enough as to require realignment, realignment can be done by jacking up one side of the forms
until it is plumb.

Tolerances

Tolerances should be the maximum that the design and/or aes-
thetic requirements allow. Tight tolerance is expensive and should

be specified only when necessary. Tolerances specified may normally not be less than the following:

- Variation from plumb in any direction:
  1 in. (25.4 mm) in any 50 ft (15.2 m)
  2 to 3 in. (50.8 to 76.2 mm) in total building height
- Variation from grade:
  1 in. (25.4 mm)
- Variation in wall thickness:
  $-\frac{1}{4}$ in. $(-6.35$ mm)
  $+\frac{1}{2}$ in. $(+12.70$ mm)
- Relation of critical surfaces to each other:
  1 in. (25.4 mm)
- Vertical placement of blockouts, inserts, and plates:
  2 in. (50.8 mm)
- Horizontal placement of blockouts, inserts, and plates:
  $\frac{1}{2}$ in. (12.7 mm)

Takedown Slipform

The procedure for taking down the slipform once the pouring is finished takes 2 to 4 weeks. The forms are usually bolted directly to the walls and then the decks and sheathing are removed and lowered with a crane.

8.1.6   Economic Consideration in Using Slipform

There are many factors that should be considered when decide to use slipform. The building has to be designed with slipform, or a similar forming system, namely, jump forming, in mind. The core has to be repetitive on all floors and fairly simple in design. Slipforming is not economical for buildings under 250 to 300 ft (76.2 to 91.4 m) in core height, because the high initial fixed cost of setup and take-down are not overcome by the low cost of forming such a relatively short height.

The cost of slipforming concrete is very sensitive to building cross section, height, steel per yard or meter of concrete, embedments, quantity of concrete per foot or meter of height, labor policies, weather, and so on. Obviously, the cost of formwork will remain the same regardless of whether the slipform structure is 50 or 500 ft (15.2 to 152.4 m) high or whether the walls are 6 or 26 in. (152.4 or 660.4 mm) thick.

Openings in the wall not only cost money to form, but they slow down the slide, thereby increasing the cost per volume of concrete proportionally. Similarly, a basic crew of approximately 10 is required whether 10 or 20 yd^3 (7.6 or 15.3 m^3) of concrete are being placed per hour. Heavy steel, in addition to the increased cost of placing the steel, results in a reduced rate of slide and, therefore, greater concrete cost.

If the weather is cold, the concrete will set more slowly, the rate of slide will decrease, and the cost per cubic yard or cubic meter of concrete will increase. A decision as to the economy of slipforming a structure should be based not only on the cost per cubic yard or cubic meter of concrete but also on the cost savings that may accrue as a result of a decrease in construction time.

There are many costly mistakes that can slow down the slipforming process, thereby making it uneconomical to slipform. Honeycombing is one that occurs frequently. This is usually due to poor vibration of the concrete. The misplacement of inserts and plates, or if the walls become too out of plumb, can lead to going back and ripping out sections of wall, or the redesign of the steel. One of the most costly mistakes that could happen is when concrete sets to the forms, making it impossible for the forms to move. This occurs when the concrete mix is not proper, or the rate of climb of the forms is slowed down for some reason. If this happens, the whole job is stopped and the set concrete is ripped out of the forms. Many times all the sheathing must be replaced. When this occurs on a job, it usually will set the schedule back 4 to 6 weeks, making slipforming an unprofitable venture.

Also, the mere nature of slipforming requires a more experienced crew and superintendent because the process is more difficult to control. Because of complex design of the core of the slipform, design changes are not easily forgiven. Once the slip is started, if a change in design is needed, even if the change is near the top of the building, the slip must be stopped in order to allow time for the redesign.

8.2 SELF-RAISING FORMWORK SYSTEM

8.2.1 Self-Raising Forms

In the last decade self-raising forms have become more popular in construction involving repetitive vertical forming. Their popularity may be attributed to the fact that they raise themselves, without crane assistance, in sequence with the casting operations. This crane independent system is very efficient in congested sites. Therefore crane costs and construction time are reduced. In addition, the concrete finish resulting from using self-raising forms is of higher quality than the one produced by slipform.

Self-raising forms have not been limited to the construction of exterior building walls. They have also been used to construct building columns, dam faces, nuclear reactor containment, cooling towers, bridge piers and elevator shafts. Along with special attached form liners, a wide range of different architectural finishes or textures can be obtained.

8.2.2 Self-Raising Form Components

The components of the self-raising forms may be customized to meet the different loads and geometry of different structures. Regardless of the system, basic similarities do exist.

The Lifting Towers

The lifting towers or frames are usually two to three stories high, in which the bottom supports of the tower are anchored to the previous lift of concrete. Attached to the lifting tower are hydraulic rams or jacks that pull the form system to the next level, raising all forming and platform areas. (See Figure 8.2.)

Forms

The form panels, often referred to as gang form panels, are made with aluminum framing members and large plywood sheets for a smooth finish. The form panels are anchored into previous cast concrete lift below. For an architectural finish, form liners may be used when finishing is required. It should be noted that the advantage of using self-raising form over slipform is its ability to produce architectural concrete by using form liners.

Rams

The raising rams are typically pulled up by either hydraulic, air powered, or electric power. Each type has its advantages and disadvantages, depending on site conditions, design loads, and the designer's preference. The designer must locate the ram lifters at suitable points on the grid so that loads will be distributed properly over the system. The rams are capable of carrying high loads; as a result, relatively few are needed. Also, each ram is self-contained, allowing for individual control of each raise and also reducing the maze of hydraulic hoses or electric cords needed.

Stripping Jacks

Self-raising form systems have different methods of form stripping. The system in Figure 8.2 uses a roll-back system in which the form is mounted on a trolley system that allows it to be stripped, rolled back and then moved to the next level. Another type of self-raising form uses stripping jacks to remove the forms away from the finished concrete and also to pull the form raiser away from the concrete, allowing it to move to the next anchoring location by the

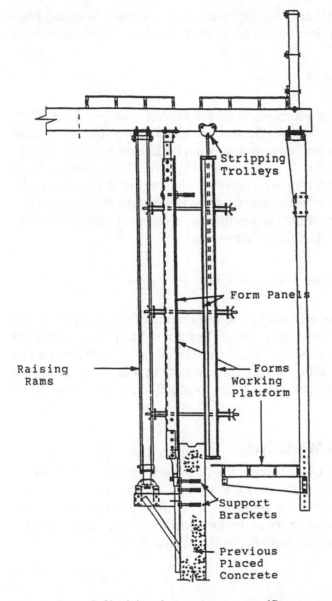

Stripping
Trolleys

Form Panels

Raising
Rams

Forms
Working
Platform

Support
Brackets

Previous
Placed
Concrete

Figure 8.2 Self-raising form component. (Courtesy of Patent Scaffolding Co.)

raising rams. The stripping jacks are then reversed, bringing both form and form raiser to the anchoring location. (See Figure 8.2.)

Working Platform

Work platforms are located on all self-raising systems. They allow workers comfortable and efficient working space to place steel and concrete, and to bolt or unbolt forms at different heights. The installation of guard rails and toe boards provide increased safety at great heights, which may contribute to worker productivity. All platforms travel with the forms, thus eliminating the need to dismantle and reassemble the platform at every work cycle. (See Figure 8.2 for work platform illustration.)

Support Brackets

Support brackets are attached to the bottom of the inner form and to the bottom of the raising ram. They are subsequently anchored to the reinforced concrete structure (see Figure 8.2.) When raising either form or lifting tower, the support brackets are detached from the structure, allowing them to travel to the next level to be anchored.

8.2.3   Typical Work Cycle

Before the work cycle begins, the formwork designer must determine the needs specific to the structure, so that design, working drawings, and other conventional form design requirements are completed. Upon delivery of the formwork system, a forming crew of six to ten workers may begin the typical work cycle.

Starter Wall

The self-raising forms can be initially placed on the foundation mat, but the process works better when a starter wall that takes the shape of the wall foot print has initially been cast. The starter wall should be about 6 ft (1.83 m) tall and have all the necessary anchorage in place, ready to support the forms and rams for the next level of casting.

Second Lift

Before the second lift can begin, the necessary anchors must be in place and the first lift must be of sufficient strength. Tower cranes are used to assemble and place the form system on the starter wall.

Form and Ram Placement

The inner form and raising rams are attached to the anchors when the first lift has gained sufficient strength. Once the inner form is in place, the outside form is rolled back on the trolley system, permitting the cleaning and oiling of forms, installation of form liners, steel reinforcement, embedments, and box-outs. The outside forms are then rolled into position, allowing the installation of taper ties.

Concrete Placement

Once the forms and ties have been inspected and oiled, the forms are ready for concrete. Minimum concrete strength must be attained, usually through cylinder strength tests, before the forms can be raised.

Stripping

Once the concrete in the second lift has attained sufficient strength, the taper ties and other embedded items are removed. The outside forms are then rolled back on the mounted trolleys.

The interior forms, which are attached to the top grid beam, allow for small movement and are released at this time.

The form support bracket at the bottom is also released at this time, leaving the system supported on the rams and the ram support brackets.

Third Lift

The system is ready to be raised to the next level once the system is supported on the ram and the ram support brackets. A typical work cycle can be seen in Figure 8.3.

Form Raising

Through a central console, the entire form is raised by the hydraulic rams to the next elevation. Individual ram controls used to do the fine grading, levels the entire platform, adjusting the

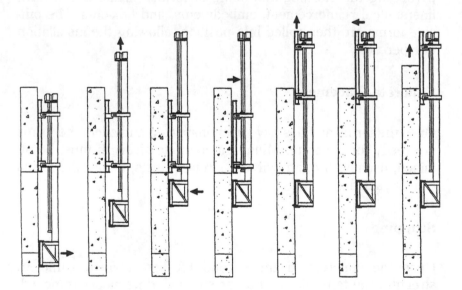

Figure 8.3 Typical work cycle.

plumbness of the core. The necessary cleaning and oiling of the interior form is done at this time.

Form Placement

The form support brackets are attached to the anchors installed in the concrete near the top of the previous lift. With the outside forms rolled back, installation of rebar, embedments, and box-outs can proceed. Once completed, the outside forms are rolled into position, and the taper ties are installed ready for the next pour. The hydraulic ram usually remains extended and attached to the second lift below during the casting of concrete.

Ram Placement

Once the concrete has been cast, the ram support bracket is detached from the anchors in the first lift, and the ram is retracted. After retraction, the ram is then reattached to anchors in the lift above.

Stripping

Once the concrete has gained sufficient strength, the stripping process is once again done.

Takedown Form

The work cycle is repeated until the desired height is reached. When the form reaches the top of the building it is normally disassembled in units and lowered by tower crane, starting with the working scaffold and the outside roll-back forms. Anchorage inserts hold the inner forms and rams, while the upper beams and platforms are disassembled and lowered. Finally the inner forms and rams are detached and lowered.

8.2.4 Advantages

Cost

The cost for any formwork system is important in controlling the cost of the project. The initial cost of purchasing or renting a formwork system and the number of reuses may determine which system to use. The initial costs for self-raising forms are typically in the range of \$1 to \$2 million a set. A set of forms are able to pour one complete floor every two to three days. From this initial cost, three-quarters of the cost is toward the purchase of the forms, while the remaining quarter is used to rent the raising equipment. To make the purchase and rental of these forms feasible, the project must have a minimum of 20 floors. Typically a set of forms can be reused for a minimum of 55 floors with little rehabilitation to the forms. The raising equipment on the other hand can be used for three to four jobs with only the simple replacement of bolts and nuts before each new job. Therefore, with the number of reuses in both the forms and raising equipment, the self-raising forms can be used from job to job for increased savings.

Construction Time

The speed of construction has assumed an increasing importance in reducing the overall cost of a building project. The rapid and efficient method of stripping, raising, and positioning of formwork through the self-raising forms has increased the rate of construction of the service core and lift shafts. Traditional forming techniques typically take four to five days to complete a 30,000 ft^2 (2800 m^2) floor, but with self-raising forms only two to three days are needed to complete the same floor. The stripping and raising of the forms from one floor to the next typically takes 20 to 30 minutes. This efficient method of forming results in direct cost reduction is due to reductions in construction time.

Labor Force

The labor force needed for the stripping, raising, and positioning of the forms is greatly reduced from the conventional formwork systems. Typically 10 to 12 workers are needed as compared to 20 or more for conventional or jump forming. The work crews are split into groups of two to three workers, which are then assigned a particular task. The workers usually work in a clockwise manner around the building with one crew working ahead of the others. The self-raising forms are prefabricated and remain connected until the top of the building is reached. This reduces the need to assemble and disassemble the forms after each lift.

With the use of built in hydraulic rams to raise the formwork system, the requirements for crane time and crane attendants is greatly reduced. Craning operations are only needed to initially place the formwork system and to disassemble the system once the forming is complete. This allows the crane to be used elsewhere on the job site, thus possibly increasing productivity elsewhere since the crane is rarely needed with the self-raising forms.

Versatility

The self-raising formwork system has many characteristics that make it versatile in the construction industry.

- The forms may be designed for almost any plan, shape, and size structure.
- The forms can be easily adjusted during construction to accommodate changes in wall thickness.
- The application of form liners provides numerous architectural finishes.
- Forms can be reused approximately 50 times or more.
- The formwork system is preassembled, therefore make-up area is not necessary.

Quality

The quality of the self-raising forms may be attributed to the fact that they are preassembled in a factory-type system. This eliminates the uncertainty that often surrounds equipment fabricated on site. The factory setting allows the formwork to be assembled before it is needed, allowing proper quality assurance programs to be implemented.

Safety

Safety on the construction site plays an important role in the productivity of the project. It should begin in the planning and management of a project continuing on down to the equipment and the workers. The self-raising forms have a number of safety features.

- The self-raising forms are preassembled in a factory-type setting, which provides predictable strengths and quality assurance of the system.
- The working platforms on the formwork system include guard rails and toe boards, thus improving safety and productivity at great heights.
- A safety factor of 3 is usually used when designing hydraulic rams.
- Forms not dismantled between raises result in fewer injuries from stripping.
- The raising of the forms are completely controlled by workers at the form, instead of crane operators and crane attendants a distance away. Therefore, self-raising forms can be stopped almost immediately in case of emergency.

8.2.5   Limitations

With the advancement of technology in the construction field, limitations of a system or technique are often reduced. The result is increased efficiency in costs and time, versatility, quality, and safety. Every formwork system, including the self-raising forms, has its limitations and range of uses.

- Members of the manufacturer's staff experienced in operating the forms are needed for several weeks for field service and to train workers, which may introduce added costs.
- Site must be fairly accessible since forms are preassembled and may be large in size.
- An accurate wall footprint must be built as a starter wall using conventional wall forming.
- The initial cost of the self-raising system is much larger than other formwork systems.
- The formwork system is economically used only for structures 20 stories or higher.
- Changes in wall size and/or location during construction are expensive and impact schedule.

9

Selection Criteria for Vertical Formwork System

9

Selection Criteria for Vertical Formwork System

9.1 ... affecting the Selection of Vertical Formwork System 246

9.2 ... the Vertical Formwork System using the Comparison Table ....

A large portion of the cost of the structural frame is the cost of formwork for columns and walls. Typically, the selection of a formwork system is made by a senior member of the contractor's organization or by a consultant from a formwork company. The decision is heavily based on that individual's experience. This chapter presents a criteria for selecting vertical formwork system for buildings. The chapter is concluded by presenting a table to assist the contractor in selecting the appropriate vertical formwork system for walls and columns.

9.1 FACTORS AFFECTING THE SELECTION OF VERTICAL FORMWORK SYSTEM

Vertical supporting structures such as buildings' cores in high-rise construction, towers, or silos are typically a critical activity that control the pace of the project progress. Consequently, when selecting a core forming system, time is a critical factor. Duration that must be considered are the movement of the form from floor to floor, original assembly, time to set rebar and inserts within the form, stripping time, close-in time, and final disassembly.

Other factors to consider in formwork selection are the amount of labor required to strip, set, pour, and control the system; the amount of precision needed as far as plumbness and corner tolerances, ease of lifting, and the designer's intent when developing the structural system. Other methods substitute the manual labor with valuable crane time. The decision between labor or crane time requires careful financial analysis. As explained in Chapter 8, crane-independent systems have been developed that do not require crane time and require considerably less labor than other systems. However, these systems are generally proprietary and require a significant investment.

Precision requirements make some systems better than others. On huge towers, cores must remain very plumb due to elevator tolerance requirements. The formwork must have a method of remaining plumb and level. If the formwork is moved piece by piece, each piece must be checked for being plumb and level, which leads to a gross amount of field engineering. Most systems become increasingly more difficult to keep plumb and level as wear and tear loosens corners and the form deteriorates in general. Also, wind loads at higher elevations tend to deform the system.

The architect usually doesn't design a building with any particular form system in mind. In some cases, such as slipforming, the design must reflect the method of forming. However, it is usually assumed the building will be conventionally formed.

Many other factors that affect the selection of vertical formwork systems for buildings are similar to those factors affecting the selection of horizontal formwork systems. However, there are some factors that are particularly important to the selection of vertical formwork systems. Among these are:

1. Factors related to building architectural and structural design, including lateral supporting system(s) and building shape and size
2. Factors related to project (job) specification, including concrete appearance and speed of construction
3. Factors related to local conditions, including area practices, weather conditions, and site characteristics
4. Factors related to the supporting organizations, including available capital, hoisting equipment, home-office support, and availability of local or regional yard supporting facilities

An overview of all the factors affecting the selection of vertical formwork systems is shown in Figure 5.1 that included the factors that are critical to selection of both horizontal and vertical formwork systems. The following sections are focused on those factors that are related solely to the selection of vertical formwork systems.

9.1.1 Building Design: Lateral Load Supporting Systems

Buildings are classified as being tall when their height is between two and three times as great as their breadth. For example, a building with a minimum dimension of 50 ft (15.2 m) in plan, and a height of 100 ft (30.5 m) or more is considered a high-rise building.

One of the major characteristics of high-rise building design is the need to resist the lateral forces due to winds, earthquakes, and any other horizontal forces. As a result special structural elements must be provided to resist lateral forces and prevent or minimize the building sway. In the following sections, a brief description of lateral load resisting systems, along with their corresponding height limitations are presented in Figure 9.1 and Table 9.1.

Type I Structure (Rigid Frame System)

Rigid frame systems consist of rectangles of vertical columns and horizontal beams connected together in the same plane. It should be noted that a bearing wall is a special case of the rigid frame system.

Type II Structure (Shear Walls)

The shear wall formation is a thin slender beam cantilevered vertically to resist lateral forces. It can take the form of a rectangle or box-shaped core, which can be used to gather vertical transportation and energy distribution systems (e.g., stairs, elevators, toilets, mechanical shafts).

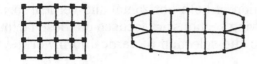

Rigid Frame System (a)

Shear Walls (b)

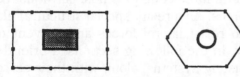

Framed Shear Wall (c)

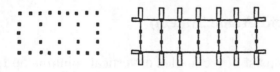

Framed Tube (d)

Tube-in-Tube (e)

Figure 9.1   Lateral loads structural supporting system.

Table 9.1 Structural Systems and
Building Height

Structural system	Optimum height
Rigid frame	Up to 20 stories
Shear wall	Up to 35 stories
Framed shear wall	Up to 50 stories
Framed tube	Up to 55 stories
Tube in tube	Up to 65 stories

Type III Structure (Framed-Shear Wall Systems)

This system consists of a combination of frames which utilize beams and columns with the shear wall designed to resist lateral loads.

Type IV Structure (Framed Tube)

This tube is a structural system in which the perimeter of the building, consisting of vertical, closely spaced supports, connected by beams or bracing members, acts as a giant vertical internally stiffened tube.

Type V Structure (Tube-in-Tube Systems)

This system is a combination of the framed-shear wall (type III structure) which acts as an interior tube, and the framed tube (type IV structure), which acts as an exterior tube. The floor structure ties the interior and exterior tubes together to allow them to act as a unit to resist lateral loads.

Table 9.1 indicates the optimum height of each lateral support system. It should be noted that Table 9.1 serves only as a rough guide for determining the suitable lateral supporting system for each building height. Judgment, experience, and the personal preference of the designer are all important factors in choosing the proper lateral support system.

9.1.2 Building Design: Building Shape

A building's structural layout is either uniform "modular" or irregular. Uniform "modular" design is characterized by regular spacing between columns and walls, equal story heights, and regular spacing of cantilevers and openings. Irregular design is characterized by irregular positions of the different structural elements and broken lines or irregular curves in architectural plans.

9.1.3 Job Specifications

Contract documents always include information about the quality of the finished concrete and the time needed to finish the project. A description about how this information can affect the selection of the formwork system is provided below.

Concrete Finish

Surface quality and appearance are always referred to as the concrete finish. Concrete surfaces can be classified as rough finish "as-cast surface," smooth exposed, covered (with a special cladding material or painted), or textured (with an architectural surface texture and treatment).

As-cast concrete finish typically shows some irregularity on the surfaces and may contain some concrete surface defects. An as-cast concrete finish is usually found in concrete buildings where appearance is not important, such as warehouses or silos.

Exposed concrete finish is characterized by smooth regular concrete surfaces, and regular positions of form ties. Exposed concrete surfaces are typically found in columns and bearing walls.

Architectural concrete is favored in vertical concrete elements (columns and walls). As a result, concrete finish must be considered as one of the major factors in the selection of a formwork system. Architectural concrete requires a careful selection of a formwork system which includes stiffer form liners, tighter joints, smoother finishes, and more care in implementing chamfers and justifications.

Speed of Construction

The most important advantage of using a sophisticated formwork system is the speed of construction. The speed of construction affects cost because it determines the time when the building will be available for use and also reduces the financial charges. The major factors that determine the speed of construction follow.

Floor Cycle. Faster floor cycle is always desirable for contractors and owners. For contractors, faster floor cycle allow the contractor to finish on schedule or earlier which reduces the overhead cost. For the owner, faster floor cycle reduces his/her short-term financial charges and allows early utilization of the constructed facility.

The use of efficient vertical formwork systems such as slip formwork and self-raising formwork allow the contractor to complete the casting of one story every two to three days. Also, vertical formwork systems control the pace of progress for horizontal concrete work.

Rate of Placement. The speed at which concrete is placed in vertical formwork has the largest influence on the lateral pressure that is imposed on the formwork. The lateral pressure is increased by increasing the rate of placement, up to a limit equal to the full fluid pressure. Higher rates of placement influence the size, material, and tie pattern of the selected system.

Construction Sequence. Tall buildings typically have an inside core to resist lateral loads. One alternative is to completely construct the inside core in order to create a "closed" area for the other trades to start. This alternative has proven to be faster than constructing and finishing floor by floor. This can be accomplished by using slipforms or self-raising forms.

9.1.4   Local Conditions

The nature of the job, including local conditions, is one of the primary factors in formwork selection. Some of the factors that should be considered are explained below.

Area Practice

Area practice has an important impact on the selected formwork system. This influence is evident in slipform operations which are highly labor intensive and usually subject to premium pay. Research indicates that slipforms are most prevalent in the northeast, southeast, and Hawaii.

Weather Conditions

Vertical forming systems are sensitive to weather conditions. Typically, in vertical forming systems, the newly placed concrete is supported by the wall already cast below it. The lower wall section must gain sufficient strength to support the fresh concrete above. The rate of strength gain for the lower wall is influenced by ambient temperature, moisture content, and freezing and thawing cycles.

Another factor that affects the economy of the selected system is the effect of stopping forming and concreting because of extreme weather conditions. For example, in slipforming, the work is usually continuous, 24 hours around the clock, with a minimum crew requirement of 50 to 75 laborers for a medium size shear wall. If the slipforming stops because of the weather conditions, the contractor has to pay the workers a show-up time, plus the cost of inactive cranes and their operators. As a result, if severe weather conditions are expected, some vertical formwork systems, such as slip forms, should be avoided.

Site Characteristics

Construction sites are generally classified into downtown restricted sites and open, suburban, or unrestricted site condition.
Gang and jump formwork require good crane service. As a result,
it is difficult to use these formwork systems in restricted site conditions. On the other hand, slipform and self-raising formwork are
crane-independent systems and can be used in restricted site conditions.

9.1.5   Supporting Organization

As previously indicated in Chapter 5, most of the ganged formwork
systems (i.e., jump form, slipform, and self-raising formwork) require high initial investment and intensive crane involvement.
However, high repetitive reuse can make these systems economically competitive. Availability of capital investment is a must for
utilizing these systems.

Hoisting Equipment (Cranes)

An important factor that influences the selection of the formwork
system is the availability of crane time. Crane time is defined as
the time in which the crane is engaged in raising and lowering
construction materials and tools. In congested site conditions
where installing more than one crane is difficult, the limited crane
time available for formwork erection to meet the project completion date becomes a major factor that may lead the formwork designer to choose crane-independent systems such as self-raising
formwork or slipform alternatives (which requires no crane time).

Home-Office Support

When deciding to use a special forming technique, the contractor
has to evaluate his or her own in-house expertise, which includes

trouble-shooting experience and safety management. For example, in vertical forming systems such as slipform and self-raising
forms, the in-house experts have to deal with special problems
such as leaking hydraulic equipment, leveling of the hoist jacks,
keeping the forms plumb within specified tolerance, and placing
inserts and openings under the fast rate of placement.

Safety management is another area of in-house expertise that
should be available to support a specific forming technique. For
example, the availability of fire protection expertise is necessary
in slipforming to prevent a fire several hundred feet in the air resulting from the flammable oil used in the hydraulic jacks. The
availability of such expertise may be a factor which determines if
a special forming technique is or is not used.

9.2   CHOOSING THE PROPER FORMWORK
SYSTEM USING THE COMPARISON TABLES

Table 9.2 is presented to help the formwork designer/selector
choose the appropriate vertical formwork system. These tables
show the relationship between the factors affecting the selection
of formwork systems and the different forming systems available
for vertical concrete work. The user must first list all the known
major components of the project and then compare them to the
characteristics listed in the table under each forming system. The
best formwork system can then be identified when the project features agree with most of the characteristics of a particular system.
These tables can also be used by architects to make some minor
adjustments in their design to accommodate the use of an efficient
formwork system.

9.2.1   Example Project

The Tabor Center is a 1.3 million ft^2 (120,000 m^2) [20,000 ft^2 (1860
m^2) per floor] multi-use facility in the center of downtown Denver.
It consists of twin office towers of 32 and 40 stories. The tower
structure is composed of three elements: the exterior wall, the cen-

Table 9.2 Factors Affecting Selection of Vertical Formwork Systems

		Formwork systems / Influence factor	Conventional column/ wall form	Ganged forms • load gathering system	Jump form	Slip form	Self-raising form
Building design	Lateral support	Lateral support system	Most suited for frames and retaining walls	Shear walls, Bearing walls, Retaining walls	Shear walls, Frames and framed shear wall	Shear walls	Shear Walls, Tube systems, Tube in tube
	Building shape	Building height	Up to 120 ft	Up to 350 ft	Up to 350 ft	Average 400 ft, Min recorded = 60 ft, Max recorded = 600 ft	At least 300 ft. No max.
		Column /wall size and location	System can handle variation of wall size and location	System can handle variation of column/	System can handle moderate variation of columns/walls size and location	Walls should be of the same location, Walls size variation can be accommodated	System can handle resonably modular design
		Openings/ inserts	System can handle openings/inserts of different size and location	Variation in opening's size and location can be accommodated at additional cost	Openings/inserts should be regually occuring from floor to floor	Should be minimum, Too many openings/ inserts make this system impractical	System can handle moderate variation in openings size and location
		Concrete finish	"As-cast" concrete finish	Produces smooth exposed concrete finish, Tie pattern and number should be designed, Form liners can be used to produce architectural concrete	Produces smooth concrete finish	System produces rough concrete finish, No ties	Smooth concrete finish, Form liners can be used
Job specification	Speed of construction	Construction sequence	Slabs and walls are placed concurrently	Slabs and walls are placed concurrently, Walls can be placed ahead of the floor slab	System is used when no floor slab is available	Typically, walls are placed entirely or at least several stories ahead of the floor	Walls are ahead of the floor. Other method is used for the first 2–3 stories
		cycle Time	1 Floor/week	1 floor every 3–4 days	1 floor every 2–3 days	1 floor every day, Rate of placing=8–20 in./hr	1 Floor every 2–3 days

Table 9.2 Continued

Formwok system	Influence factor	Conventional column/wall form	Ganged forms + load gathering system	Jump form	Slip form	Self-raising form
Local conditions — Site characteristics	Area practice	More efficient in areas of high quality, low-cost labor force	Work best in high-cost, low-quality labor force	System is easy to learn and adapt Learning curve is quite short	System can be learned in 2–3 weeks	System requires high-quality supervision
	Weather	Generally not a major factor	A major factor, walls should have sufficient strength before stripping which is largely influenced by weather condition High winds limit the crane movement		Hot or cold weather affects the concrete rate of setting which slow the rate of rise	In cold weather, forms should be protected and concrete should be heated
	Access to site	Generally not a factor for loose forms	Can be a major factor if the system is pre-assembled in a local yard facility	Site must be accessible, forms can be up to 16 ft high and 44 ft wide	System must have a limited access for material delivery	Min. site access is required for concrete placing and rebar delivery
	Site size	Not a factor	Can be a major factor if the forms have to be built in site	Not a major factor, forms are preassembled and unloaded directly	Not a major factor, system can be used in restricte./ small sites	

Table 9.2 Continued

Influence factor	Formwork systems	Conventional column/wall form	Ganged forms • load gathering system	Jump form	Slip form	Self-raising form
Cost	Stripping	Hand strip High stripping cost	Crane is used to strip the system High stripping cost	Forms are equiped with mechanism for stripping Min stripping cost	Forms are stripped at the end of the project Min. stripping cost	Forms are equiped with mechanism for stripping Min stripping cost
Cost	Reuse	Less than 10	Between 40 and 50 Reuses could be horizontally or vertically	Between 15 and 30	Between 50 and 100 (i.e., between 200 and 400 ft high)	At least 30 reuses should be available vertically
	Location of adjacent building and obstruction	Generally not a factor	A major factor, system must have a free space to be moved from floor to floor		Minimum free space should be available for crane movement	Not a major factor, system can be used in downtown restricted areas
Hoisting equipment	Crane time	Not a factor, system car be hand-set	Crane-dependent system, crane time is a must	System substantially reduces crane time Average crane time pick = 20 min.	Crane is used only for materials delivery and concrete placing	Crane-independent system
Hoisting equipment	Operating system	Hand-set system, crane increases system efficiency and reduces cost	Crane-set system Crane serves two functions: lifting and supporting the forms	Crane is used only to lift the forms Crane is not used for forms dismantling	Locomotion is provided by electric, pneumatic,or hydraulic jacks climbing on smooth steel rods	System is lifted by hydraulic, electric, or pneumatic lifters.
Safety mangt.	Safety	No special safety features is required	Special care for handling the large ganged units by crane	Safety features Safe guarded platform No one needs to be on the form during crane handling	For hydraulic systems, special safety precautions must be taken to prevent fire several hundred feet above the ground	
Yard facility	Supporting yard facility, supplier or make-up area	Not a major factor, but system is more efficient, if a local yard facility is available	A major factor, system must have an adequate make-up area or close by supplier	System is rented or purchased	Continuous materials delivery is a must. Un-interrupted concrete placement must be assured.	System is pre-assembled, make-up area is not a factor. Local supplier must be available

Supporting organization

tral core, and the interior deck area that ties the two together (tube in tube). The core is structural steel with a concrete diaphragm or infill walls up to floor 12 to resist part of the lateral loading. The deck area is also structural steel with a metal deck and concrete fill. The exterior wall consists of closely spaced columns (tube system). A study showed that the crane time is not adequate for hoisting formwork and it will only be used for material handling and concrete placing. Architectural concrete is required for the exterior walls.

System Selection Via Tables

Table 9.2 was used to select the formwork system for the exterior "wall." The selection was a self-raising form for the following reasons:

1. The self-raising form is suitable for buildings higher than 25 stories.
2. It can accommodate architectural concrete requirements.
3. The self-raising form is a crane-independent system which is suitable for restricted site conditions.

This case study was extracted from an article entitled "Building Production and Quality into Architectural Concrete." The whole article was devoted to the problem of selecting the formwork system. The author is an expert in selecting formwork systems and he attributed his selection to the same reasons stated above. In his article, he explained that renting or buying another crane was impossible because of the site conditions in "downtown Denver." It was also economically unfeasible.

References

Horizontal Forms

W. R. Anthony, Forming Economical Concrete Buildings—Proceedings of the Third International Conference, SP-107, American Concrete Institute, 1988, pp. 1–28.

Constructional Review (North Sydney), Vol. 56, No. 2, May 1983, pp. 17–27.

V. H. Perry and S. K. Malhotra, *International Journal for Housing Science and Its Applications*, Vol. 9, No. 2, 1985, pp. 131–140.

Stephen Timpson and James M. Henry, Forming Economical Concrete Buildings—Proceedings of the Second International Conference, SP-90, American Concrete Institute, 1986, pp. 201–218.

Antonio Tramontin, *L'Industria Italiana del Cemento (Rome)*, Vol. 55, No. 585, January 1985, pp. 20–31.

Conventional Forms

J. Bullock, Forming Economical Concrete Buildings—Proceedings of the Third International Conference, SP-107, American Concrete Institute, 1988, pp. 121–136.

Concrete Construction, Vol. 30, No. 2, February 1985, pp. 197–200.

Concrete Products, Vol. 87, No. 7, July 1984, pp. 41, 45.

Engineering News-Record, Vol. 210, No. 10, January 27, 1982, pp. 24–25.

Engineering News-Record, Vol. 219, No. 10, September 3, 1987, p. 13.

Mark Fintel and S. K. Ghosh, *Concrete International: Design and Construction*, Vol. 5, No. 2, February 1983, pp. 21–34.

Denis A. Jensen, *Concrete Construction*, Vol. 26, No. 11, November 1981, pp. 883–887.

James E. McDonald, *Concrete International: Design and Construction*, Vol. 10, No. 6, June 1988, pp. 31–37.

J. Schlaich and W. Sobek, *Concrete International: Design and Construction*, Vol. 8, No. 1, January 1986, pp. 41–45.

Shoring Forms

Xila Liu, Wai-Fah Chen, and Mark D. Bowman, *Journal of Structural Engineering—ASCE*, Vol. 111, No. 5, May 1985, pp. 1019–1036.

Pericles C. Stivaros and Grant T. Halvorsen, *Concrete International*, Vol. 14, No. 8, August 1992, pp. 27–32.

Flying Forms

Kim Basham and Jeff Groom, Concrete Construction, Vol. 33, No. 3, March 1988, pp. 299, 301.

Bill Blaha, *Concrete Products*, Vol. 88, No. 1, January 1985, pp. 34–35.

Concrete Construction, Vol. 29, No. 11, November 1984, pp. 949–953.

Engineering News-Record, Vol. 208, No. 10, March 11, 1982, pp. 30–31

Engineering News-Record, Vol. 210, No. 10, January 27, 1982, pp. 24–25.

Craig G. Huntington and Y. C. Yang, *Civil Engineering—ACE*, Vol. 53, No. 7, July 1983, pp. 39–41.

Tunnel Forms

Bill Blaha, *Concrete Products*, Vol. 86, No. 9, September 1983, pp. 21–25.

Housley Carr, *Engineering News-Record*, Vol. 216, No. 22, May 29, 1986, pp. 37–38.

Civil Engineering—ASCE, Vol. 53, No. 11, November 1983, pp. 60–63.

Constructor, Vol. 67, No. 8, August 1985, pp. 23–25.

Engineering News-Record, Vol. 215, No. 24, December 12, 1985, pp. 25–27.

Engineering News-Record, Vol. 205, No. 18, October 29, 1981, p. 16.

R. Harrell, *Nondestructive Testing*, SP-112, American Concrete Institute, 1988, pp. 153–164.

D. Sabine and E. Skelton, Proceedings, Institution of Civil Engineers (London), Part 1, Vol. 78, December 1985, pp. 1261–1279.

Anne Smith, *Concrete Construction*, Vol. 36, No. 8, August 1991, pp. 592–593.

Lorraine Smith, *Engineering News-Record*, Vol. 216, No. 14, April 3, 1986, p. 36EC.

Vertical Forms

W. R. Anthony, Forming Economical Concrete Buildings—Proceedings of the Third International Conference, SP-107, American Concrete Institute, 1988, pp. 1–28.

Phillip J. Arnold, *Concrete Products*, Vol. 89, No. 7, July 1986, pp. 15–18, 42.

Concrete Construction, Vol. 27, No. 2, February 1982, pp. 173–180.

M. S. Fletcher, *Concrete (London)*, Vol. 16, No. 5, May 1982, pp. 41–44.

N. J. Gardner, Proceedings, Institution of Civil Engineers (London), Part I, Vol. 80, February 1986, pp. 145–159.

N. J. Gardner, *ACI Journal*, Proceedings Vol. 82, No. 5, September–October 1985, pp. 744–753.

N. J. Gardner, *Concrete International: Design and Construction*, Vol. 6, No. 10, October 1984, pp. 50–55.

Awad S. Hanna and Victor E. Sanvido, *Concrete International: Design and Construction*, Vol. 12, No. 4, April 1990, pp. 26–32.

T. A. Harrison, *Concrete International: Design and Construction*, Vol. 5, No. 12, December 1983, pp. 23–28.

Bruce A. Lamberton, *Concrete International: Design and Construction*, Vol. 11, No. 12, December 1989, pp. 58–67.

Paul H. Sommers, *Concrete Construction*, Vol. 29, No. 4, April 1984, pp. 392–394.

Gang Forms

Constructioneer, Vol. 41, No. 23, December 7, 1987, pp. 22–23.

Jerome H. Ford, *Concrete Construction*, Vol. 27, No. 1, January 1982, pp. 51–57.

Donald Schaap, *Concrete International: Design and Construction*, Vol. 11, No. 2, February 1989, pp. 37–41.

Charles Steele, *Concrete (Chicago)*, Vol. 46, No. 12, April 1983, pp. 18–21.

Jump Forms

ACI Committee 313, *ACI Structural Journal*, Vol. 88, No. 1, January–February 1991, pp. 113–114.

Constructional Review (North Sydney), Vol. 56, No. 4, November 1983, pp. 24–29.

Slip Forms

ACI Committee 311, Special Publication No. 2, 7th Ed., American Concrete Institute, Detroit, 1981, 400 pp., $24.95 ($19.95 to ACI members).

ACI Committee 313, *ACI Structural Journal*, Vol. 88, No. 1, January–February 1991, pp. 113–114.

ACI Committee 347, *ACI Structural Journal*, Vol. 85, No. 5, September–October 1988, pp. 530–562.

John A. Bickley, Shondeep Sarkar, and Marcel Langlois, *Concrete International*, Vol. 14, No. 8, August 1992, pp. 51–55.

Ian Burnett, *Concrete International: Design and Construction*, Vol. 11, No. 4, April 1989, pp. 17–25.

Bill Blaha, *Concrete Products*, Vol. 88, No. 3, March 1985, pp. 26–28.

Carl V. Carper, *Construction Equipment*, Vol. 71, No. 2, February 15, 1985, pp. 100–102.

Concrete Construction, Vol. 27, No. 7, July 1982, pp. 571–573.

Concrete Construction, Vol. 27, No. 2, February 1982, pp. 173–180.

Concrete Construction, Vol. 27, No. 1, January 1982, pp. 64–67.

Concrete Construction, Vol. 29, No. 11, November 1984, pp. 949–953.

Concrete Construction, Vol. 31, No. 8, August 1986, pp. 699–708.

Concrete Products, Vol. 86, No. 11, November 1983, p. 32.

Concrete Products, Vol. 86, No. 11, November 1983, p. 43.

Concrete Products, Vol. 89, No. 4, April 1986, p. 5.

Concrete Quarterly (London), No. 129, April–July 1981, pp. 14–19.

Concrete Quarterly (London), No. 140, January–March 1984, pp. 10–11.

Constructioneer, Vol. 37, No. 24, December 19, 1983, pp. 16–17.

Constructioneer, Vol. 38, No. 7, April 2, 1984, pp. 38–39.

Lee Dennegar, *Concrete International: Design and Construction*, Vol. 5, No. 8, August 1983, pp. 30–33.

Dale Diulus, *Civil Engineering—ASCE*, Vol. 56, No. 6, June 1986, pp. 65–67.

Engineering News-Record, Vol. 207, No. 17, October 22, 1981, pp. 21–22.

Engineering News-Record, Vol. 208, No. 18, May 6, 1982, pp. 32–33.

Engineering News-Record, Vol. 209, No. 18, October 1982, pp. 28–33.

Engineering News-Record, Vol. 209, No. 6, February 11, 1982, pp. 26–27.

Engineering News-Record, Vol. 213, No. 12, September 20, 1984, pp. 61–63

Engineering News- Record, Vol. 219, No. 25, December 17, 1987, pp. 92–93

C. Hsieh and Jerome R. King, *Proceedings, ASCE*, Vol. 108, C01, March 1982, pp. 63–73.

Ralph Ironman, *Concrete Products*, Vol. 87, No. 11, November 1984, p. 24.

Hal Iyengar, *Civil Engineering—ASCE*, Vol. 55, No. 3, March 1985, pp. 46–49.

Denis A. Jensen, *Concrete Construction*, Vol. 26, No. 11, November 1981, pp. 883–887.

Maage, Lewis H. Tuthill International Symposium on Concrete and Concrete Construction, SP-104, American Concrete Institute, 1987, pp. 185–204.

Malcolm R. H. Dunstan, *Concrete International: Design and Construction*, Vol. 5, No. 3, March 1983, pp. 19–31.

Thomas O. Mineo and Catheleen M. Cassidy, *Concrete International: Design and Construction*, Vol. 5, No. 12, December 1983, pp. 44–47.

Yoichior Murakami and Tatsuo Sato, *Concrete International: Design and Construction*, Vol. 5, No. 9, September 1983, pp. 42–50.

Christopher Olson, *Building Design and Construction*, Vol. 24, No. 12, December 1983, pp. 44–47.

Charles J. Pankow, *Concrete International: Design and Construction*, Vol. 9, No. 10, October 1987, pp. 23–27.

Doug Priutt, *Concrete Construction*, Vol. 32, No. 4, April 1987, pp. 345–349.

William G. Reinhardt, *Engineering News-Record*, Vol. 217, No. 23, December 4, 1986, pp. 22–24.

Christopher E. Reseigh, *Concrete Products*, Vol. 88, No. 12, December 1985, pp. 26–29, 47.

Ernst Roeck, *Concrete International: Design and Construction*, Vol. 4, No. 6, June 1982, pp. 33–37.

Steve Steinberg, *Building Design and Construction*, Vol. 23, No. 10, October 1982, pp. 43–45.

Stephen Timpson and James M. Henry, Forming Economical Concrete
    Buildings—Proceedings of the Second International Conference,
    SP-90, American Concrete Institute, 1986, pp. 201–218.
William Zuk, *The Military Engineer* (607 Prince St., Alexandria, VA
    22314), Vol. 73, No. 474, July–August 1981, pp. 254–258.

Appendix
Bending Moment, Shear, and Deflection Equations (Metric)

Design Conditions	Support Conditions		
	1 Span	2 Spans	3 or More Spans
Bending			
Wood	$l = \dfrac{36.5}{1000}d\left(\dfrac{F_b b}{w}\right)^{1/2}$	$l = \dfrac{36.5}{1000}d\left(\dfrac{F_b b}{w}\right)^{1/2}$	$l = \dfrac{40.7}{1000}d\left(\dfrac{F_b b}{w}\right)^{1/2}$
	$l = \dfrac{89.9}{1000}\left(\dfrac{F_b b}{w}\right)^{1/2}$	$l = \dfrac{89.9}{1000}\left(\dfrac{F_b b}{w}\right)^{1/2}$	$l = \dfrac{100}{1000}\left(\dfrac{F_b b}{w}\right)^{1/2}$
Plywood	$l = 2.83\left(\dfrac{F_b KS}{w}\right)^{1/2}$	$l = 2.83\left(\dfrac{F_b KS}{w}\right)^{1/2}$	$l = 3.16\left(\dfrac{F_b KS}{w}\right)^{1/2}$
Shear			
Wood	$l = \dfrac{1.34}{1000}\dfrac{F_v A}{w} + 2d$	$l = \dfrac{1.07}{1000}\dfrac{F_v A}{w} + 2d$	$l = \dfrac{1.11}{1000}\dfrac{F_v A}{w} + 2d$
Plywood	$l = 2.00\dfrac{F_s Ib/Q}{w} + 2d$	$l = 1.60\dfrac{F_s Ib/Q}{w} + 2d$	$l = 1.67\dfrac{F_s Ib/Q}{w} + 2d$
Deflection	$l = 526\left(\dfrac{EI\Delta}{w}\right)^{1/4}$	$l = 655\left(\dfrac{EI\Delta}{w}\right)^{1/4}$	$l = 617\left(\dfrac{EI\Delta}{w}\right)^{1/4}$
If $\Delta = \sqrt{180}$	$l = \dfrac{75.1}{1000}\left(\dfrac{EI}{w}\right)^{1/3}$	$l = \dfrac{101}{1000}\left(\dfrac{EI}{w}\right)^{1/3}$	$l = \dfrac{93.0}{1000}\left(\dfrac{EI}{w}\right)^{1/3}$
If $\Delta = \sqrt{240}$	$l = \dfrac{68.5}{1000}\left(\dfrac{EI}{w}\right)^{1/3}$	$l = \dfrac{91.7}{1000}\left(\dfrac{EI}{w}\right)^{1/3}$	$l = \dfrac{84.7}{1000}\left(\dfrac{EI}{w}\right)^{1/3}$
If $\Delta = \sqrt{360}$	$l = \dfrac{59.8}{1000}\left(\dfrac{EI}{w}\right)^{1/3}$	$l = \dfrac{79.9}{1000}\left(\dfrac{EI}{w}\right)^{1/3}$	$l = \dfrac{73.8}{1000}\left(\dfrac{EI}{w}\right)^{1/3}$
Compression	f_c or $f_{c\perp} = \dfrac{P}{A}$		
Tension	$f_t = \dfrac{P}{A}$		

Notation:

l = length of span, center to center of supports (mm)

F_b = allowable unit stress in bending (kPa)

$F_s KS$ = plywood section capacity in bending (Nmm/m)

F_c = allowable unit stress in compression parallel to grain (kPa)

$F_{c\perp}$ = allowable unit stress in compression perpendicular to grain (kPa)

$F_s Ib/Q$ = plywood section capacity in rolling shear (N/m)

f_v = allowable unit stress in horizontal shear (kPa)

f_c = actual unit stress in compression parallel to grain (kPa)

$f_{c\perp}$ = actual unit stress in compression perpendicular to grain (kPa)

f_t = actual unit stress in tension (kPa)

A = area of section (mm^2)*

E = modulus of elasticity (kPa)

I = moment of inertia (mm^4)*

EI = plywood stiffness capacity (kPamm^4/m)

P = applied force (compression or tension) (N)

S = section modulus (mm^3)*

Δ = deflection (mm)

b = width of member (mm)

d = depth of member (mm)

w = uniform load per meter of span (kPa/m)

*For a rectangular member: $A = bd$, $S = bd^2/6$, $I = bd^3/12$

Index